Cambridge Computer Science Texts · 2



# Information Representation and Manipulation in a Computer

## Second edition

**E. S. PAGE and L. B. WILSON**

Computing Laboratory, University of Newcastle upon Tyne

**Cambridge University Press**

**Cambridge**
**London · New York · Melbourne**

Published by the Syndics of the Cambridge University Press
The Pitt Building, Trumpington Street, Cambridge CB2 1RP
Bentley House, 200 Euston Road, London NW1 2DB
32 East 57th Street, New York, NY 10022, USA
296 Beaconsfield Parade, Middle Park, Melbourne 3206, Australia



First published 1973
Reprinted 1975
Reprinted with minor corrections and amendments 1976
Second edition 1978

Printed in Great Britain at the
University Press, Cambridge

ISBN 0 521 22088 2 hard covers
ISBN 0 521 29357 x paperback
(ISBN 0 521 20146 2 first edition)

# Contents


| | Page |
|---|---|
| Preface | vii |
| Preface to the Second Edition | ix |
| **Chapter 1 Symbols on Paper** | |
| 1.1 Introduction | 1 |
| 1.2 Computer Operations | 2 |
| 1.3 Single Symbols | 3 |
| 1.4 Descriptions of Sounds and Positions | 5 |
| 1.5 Other Representations by Symbols | 6 |
| Examples 1 | 7 |
| **Chapter 2 Symbols and Codes** | |
| 2.1 Binary Elements | 8 |
| 2.2 External Representation | 9 |
| 2.3 Paper Tape | 9 |
| 2.4 Error Correction and Detection | 12 |
| 2.5 Coding Theory | 16 |
| 2.6 Construction of Optimal Codes | 20 |
| 2.7 Weighted Codes | 22 |
| 2.8 Punched Card Codes | 24 |
| 2.9 Bibliography | 25 |
| Examples 2 | 25 |
| **Chapter 3 Internal Representation** | |
| 3.1 Units of Storage | 32 |
| 3.2 Conversion between Scales | 33 |
| 3.3 Integers: Packed Decimal | 36 |
| 3.4 Integers: Binary Representation | 37 |
| 3.5 Fractions: Fixed Binary | 38 |
| 3.6 Floating Point Numbers | 39 |
| 3.7 Bibliography | 42 |
| Examples 3 | 43 |

Chapter 4 Information Structures 1: Arrays

4.1 Introduction 50

4.2 Storage of Arrays 53

4.3 Applications of Access Tables 59

4.4 Sparse Arrays 61

4.5 Bibliography 65

Examples 4 66

Chapter 5 Information Structures 2: Linear Lists

5.1 Introduction 74

5.2 Stacks, Queues and Deques 74

5.3 Sequential Allocation of Storage 78

5.4 Linked Allocation of Storage 85

5.5 Comparison of Sequential and linked Allocation of Storage 97

5.6 Bibliography 99

Examples 5 100

Chapter 6 Information Structures 3: Trees

6.1 Introduction and Basic Definitions 109

6.2 Traversing a Tree 115

6.3 The Transformation of Trees into Binary Trees 120

6.4 Tree Representation 127

6.5 Path Length 143

6.6 Bibliography 147

Examples 6 149

Chapter 7 Searching

7.1 Introduction 162

7.2 Scanning 163

7.3 Key Transformation (Scatter Storage) Techniques 173

7.4 Bibliography 183

Examples 7 185

Chapter 8 Sorting

8.1 Introduction 193

8.2 Internal Methods 196

8.3 Merging Methods 214

8.4 A Comparison of Sorting Methods 221
8.5 Bibliography 223
Examples 8 226

Notes on the solutions to Examples 235

Index 265

# Preface

Computer science has been one of the most rapidly growing subjects during the past decade or so, and an increasing proportion of students in universities, polytechnics and, in the later years, of some schools, now receive an introduction to computing which includes practical work in a programming language. The time has hopefully passed when this is all that could be considered necessary. A part of the other fundamental material which should, and can, be presented at an elementary stage concerns how information can be represented and manipulated. This text is intended for students on such an introductory course and requires only a knowledge of some high level programming language and the ability to write and test programs on a computer. The mathematical demands on the reader have purposely been kept as slight as possible and only in a few places, where the full analysis of some algorithm demands it, is more than elementary algebra needed. The main thread of the book will not be lost if the reader has to skip such portions.

The material in this book can be covered in about two dozen, one-hour lectures, and can be given at a quite elementary stage in a student's study of computing. We have been giving it in the first year to undergraduates intending to specialize to some extent in computing science, but it could just as easily fit into a course at a later stage for those using computing as a tool and spending a smaller proportion of their time in the subject. The text also contains little that is not required for the first stage of the British Computer Society's examination. Examples are given at the end of each of the chapters to enable the student to test his understanding of the content. Some of the examples are straightforward exercises of drill, others are rather longer questions taken from university examination papers. Such a written examination is unlikely to form the whole of the assessment for the course, normally being supplemented at least by some programming project, but these questions represent a type of assessment which is usually present at some stage. We have not indicated the year of the course in which the question has appeared because such information would only be meaningful when coupled with a knowledge of how the particular university arranged its courses. In the same way, certain questions have been taken from papers testing the work of some M. Sc. conversion courses. The third type of exercises which is included are some suggestions for rather longer programming projects which might require a month or more of a student's own time to be devoted to the subject. Some hints on the solutions of the examples are given at the end of the book, in sufficient detail, we hope, to indicate how someone who is stuck should proceed and yet sufficiently tersely to discourage a student from presenting our hints as his solution should it be set by his lecturer.

This text is short, and purposely so; it will be evident to all computer scientists from reading the general literature, and in particular the magnificent volumes of Knuth, how much work there already is on the topics touched on here. However, even specialist students of computing science will find it hard to afford the time to study the topics in the detail described by Knuth, and even harder from what we are told to afford the cost of buying the books at the outset of their study of computing science. We hope, however, that many of those following the references given here will be stimulated to study in more detail.

Several of our colleagues have assisted us by commenting on various parts of the draft of this text and we are most grateful to them. It is too much to hope that we benefitted sufficiently to eliminate all errors and for those that remain we must claim full credit. We are also pleased to acknowledge our gratitude to the Universities of Belfast, Essex, Glasgow, Leeds, Liverpool, Manchester, Newcastle, Pennsylvania, St. Andrews, Sheffield and Southampton and the Carnegie Institute of Technology for their permission to include questions from their examination papers.

E. S. Page
L. B. Wilson

# Preface to the Second Edition

The preparation of a second edition has given us an opportunity to re-write and extend certain sections and particularly to revise some algorithms whose original programming style betrayed their age. We are greatly indebted to friends and correspondents for suggestions for improvement and, to put it bluntly, for a few corrections; our special thanks are owed to Dr. Nils Andersen of the University of Copenhagen who sent many comments which have helped us. Further examples have been added together with notes on their solution.

E. S. Page
L. B. Wilson

# 1 · Symbols on Paper

## 1.1 INTRODUCTION

The good teacher or lecturer aims to present interesting and challenging ideas, results and arguments which may be quite complex in a manner which seems straightforward and simple to the student. In contrast, the composers of different forms of puzzles and brain teasers for newspapers and magazines aim to produce formulations of problems which appear intriguing and perhaps quite difficult but which still allow a simple solution once the correct method of attack has been discovered. The ways the problems are presented have a large effect upon the ease of solving them. The puzzle setter and the teacher have each made a choice of how their material should be presented - a choice guided by the effects they hope to achieve. At a more detailed level the notation chosen for the various quantities occurring in a problem can have a dominating influence on the ease with which a solution may be found. For example, some of the early representations of numbers seem to have been devised primarily for recording amounts of goods; so many cattle, so much corn, and so on. Different units had symbols of their own and other numbers were composed by the appropriate repetitions of those symbols. Such a system of representation does not place too much hindrance in the way of the operations of addition and subtraction; the symbols themselves can be repeated or deleted easily and the occasional replacement by equivalent groups of symbols performed. It is much more difficult to perform multiplication and division - operations presumably less frequently required in the applications that were routine. The same sort of difficulties occur with the more recent Roman numerals, although some additional complexities appear (and produce the material for a sequence of elementary programming examples). The hindrances to multiplication and division become much less once the cypher (a figure 0) is introduced together with the usual positional notation. For example, in the scale of ten, 3080 represents three thousands and eight tens $(3\times10^3+0\times10^2+8\times10^1+0\times10^0)$. Even here, however, some operations are performed more easily than others. In this scale multiplication by ten is an easier operation to perform than multiplication by two even, and certainly by seven. Other scales have their own properties which ease some operations and make others more difficult.

The same kind of differences in the ease of performing given operations can be noticed in many areas far removed from arithmetic or mathematics. For example, given a tuned guitar or banjo and an air described by the musical score, those with as little musical training as the authors would be at a loss to produce anything recognizable. If, however, the grid representation of the fingering is given we might manage to produce some of the right notes in the right order if not in the right tempo. Conversely, however, the fingering symbols alone would not make it easy for even a skilled musician to play the tune on a piano or on a wind instrument. This theme will constantly recur in different guises throughout this book. Any representation of information that has been chosen will govern what operations are easy and convenient to perform and, conversely, a representation should be selected taking fully into account the operations which it is necessary to perform.

## 1.2 COMPUTER OPERATIONS

The early uses to which computers were put were predominantly numerical. Ballistic and navigation tables were produced; the numerical solutions were printed for mathematical problems arising in different branches of engineering and science. Thus, at first computers tended to be regarded solely as devices for performing arithmetic operations; however, it soon came to be recognized that there was a substantial amount of logical and administrative work contained even in a computer program for numerical calculations. The contents of a storage location had to be examined to see if it was positive or negative, perhaps to control a count of how many times a loop had been obeyed or whether another iteration was needed. The contents of stores had to be moved to different locations in the machine; during the moving operation the meaning of the pattern of electrical signals would be irrelevant - a copy was being made for later interpretation. Sometimes parts of the contents of a location had to be abstracted or the contents moved relative to themselves as in a shift operation. Counts of the different sorts of operations showed that even in a numerical calculation the proportion of the administrative and non-numerical operations was high. There was thus a change in emphasis to cause computers to be regarded as devices for performing operations on symbols; numerical digits just became special cases of the more general class of all the symbols represented. We therefore shall look first at single symbols and then at groups of symbols which are used in several different applications independent of computers and shall consider the different sorts of operations that are performed upon them; later we restrict attention to uses in computers. For our purposes, therefore, we regard information as being conveyed by symbols which are distinguishable one from another and any meaning that they have will be governed by the rules of the particular context in which they appear.

## 1.3 SINGLE SYMBOLS

The most commonly used single symbols in the western world are surely the letters of the English alphabet, a, b, c, ... Notice, however, that there are many variations possible even in this simple example. In typescript letters may occur in lower case a, b, c, ... or in upper case A, B, C, ... In printing they can occur in a variety of different type founts as well; for example, italic, bold, script and many others. The letters can be different sizes - from the small print often used for the limiting conditions of guarantees and legal agreements through the sizes used for the headlines of the popular newspapers to the display characters for advertisements on the hoardings. Notice again that the choices of representation on the paper have been made in order to try to achieve one or more principal aims while satisfying to some extent subsidiary aims. For example, the 'small print' on a purchase contract can, at least charitably, be justified by the need to compress many symbols into a small space, with a subsidiary requirement that the symbols can be read with normal aids to eyesight - the microdot recording used by spies not being permitted.

Representations of characters which we recognize as the same in some respects are produced by very many different means. For example, from an ordinary typewriter upper case letters are caused by depressing a case shift key and then striking the key for the letter so that a different part of the head of the moving arm strikes the inked ribbon onto the paper. On some machines (e. g. flexowriters, teleprinters) which produce a paper tape output as well as a printed copy, a case shift key has to be struck which causes a pattern of holes on the paper tape to be produced as well as placing the machine in the state ready to print letters in upper case when the keys are struck. In some methods of printing both upper and lower case letters are represented uniquely by individual keys.

The familiar decimal digits, 0, 1, 2 . . . 9 appear on most keyboards and are adequate for representing numbers in scales of 10 or less but need to be supplemented for scales with higher radices. For example, hexadecimal numbers, i. e. those in the scale of 16, need symbols to represent the decimal 10, 11, 12, 13, 14 and 15; by one convention these have been taken to be A, B, C, D, E, F, so that the hexadecimal A5 is 165 in the scale of ten. In this case it is quite usual for some of the keys to be used for two different purposes - BED represents both a hexadecimal number and an English word meaning something to sleep in or plant flowers in. It is worth noticing, however, that we do not need to go to exotic scales of notation to find examples of double usage of symbols. Even on some common keyboards not all the decimal digits are available; for example, on many typewriter keyboards the digits zero and one are absent - the typist is supposed to produce them by using the upper case O (i. e. the letter between N and P in the alphabet) and lower case 'el', a substitution which produces errors often noticed in the output from inexpert typists.

Restrictions on the size of keyboards lead to the omission of some needed symbols and so to tricks to avoid them. Consider the mathematical signs $+ - \times \div /$ $> \geq = \neq$. The first two signs are usually available but multiplication might have to be indicated by the letters x or X - for example, on some typewriters. Other signs like $\neq$ might need to be constructed from an over-printing of $=$ by $/$; yet others like $\geq$ may have to be replaced by some combination like .GE. Some character sets require such a construction for a large number of their symbols; many of the operators of the Iverson notation in his language APL require more than one key stroke. In all these cases we notice that some choice has been made of what set of characters should be represented by a single key, how big that set should be, how other characters may be represented (if at all); a good choice has regard to the most frequent or most important uses.

## 1.4 DESCRIPTIONS OF SOUNDS AND POSITIONS

Not all phenomena are described most conveniently by alphabetical or numerical symbols for the operations which have to be performed. Descriptions by these symbols may take up too much space or too much time, or they may be less readily distinguished than a suitable stylized two-dimensional picture. A secretary taking dictation from someone speaking only moderately fluently will record the words spoken in one of the shorthand systems that have been developed. Two of the most common, Pitman's and Gregg's, represent the sounds of the words rather than their spelling. A number of basic outlines written in a few positions relative to the horizontal lines on the paper are combined to form all the words of the language. The operation of encoding by a skillful shorthand-writer can be fast enough to record what even quick speakers are saying, and the systems have surely been designed with this primarily in mind. Decoding, accompanied as it usually is by some form of transcription on a typewriter, need not be quite as quick or even in most cases quite as accurate if the context or the secretary's memory can afford some clue in the case of a slightly inaccurate written outline. The ordinary representation of a music score as well as the one quoted earlier for a banjo or guitar, uses a two-dimensional display on the printed page. In these cases the operations that need to be performed quickly are the recognition of the sounds required and what has to be done to produce them from an instrument. The converse construction, printed representation produced from the sounds, need not normally be performed at the same speed and the notation is not well adapted for this purpose. In some other spheres it is necessary to describe positions in two or three dimensions, for example, in ballet or modern dance routines. In both of these cases essentially two-dimensional forms of representation on the paper have been devised but because of the nature of the activity are perhaps more suited to recording a sequence of movements rather than to assist their execution at the desired rate.

## 1.5 OTHER REPRESENTATIONS BY SYMBOLS

The previous sections have given examples of some representations on the written or printed page and have focussed attention on properties of the representations. Paper and printing are, of course, not the only means of recording or transmitting information. The deaf and dumb alphabet, formed from positions of the hands and fingers, gives another representation of the alphabet plus a small number of words and phrases. Another visual representation of the alphabet is given by semaphore which uses the positions of flags held in the hands to describe the letters and digits. A much more extensive system for conveying messages from a previously determined list has been constructed for ships at sea which uses a variety of devices including flags of various colours. Sight is only one of the senses available for the recognition of information. Braille notation is recognized by touch, and the Morse Code in one of its varieties typically uses long or short sounds or long or short flashes of light in patterns to represent the printed symbols. Even smell gives some information, especially to animals. All these representations and the ones mentioned earlier allow some operations to be performed more easily than others. If the more awkward operations have to be performed relatively infrequently it may be worthwhile retaining the representation because it is familiar. It must, however, be recognized that a decision about the choice of a representation needs to be made, and the consequences of the choice contemplated and weighed against alternatives. The examples quoted are naturally those which have stood the test of considerable use and have proved their suitability under the most frequent types of use. We shall see later that some sort of standardization is appearing for the representation within computers of the most common characters, but whenever more unusual applications appear a choice may have to be made ab initio.

EXAMPLES 1

[1.1] For any keyboards producing a hard copy with which you are familiar, list:

(a) the single symbols in the printed representation which require more than one key stroke to produce them;

(b) those keys which represent different printed symbols according to context (as with the lower case 'el' quoted in section 1.2);

(c) those keys which produce no printing.

[1.2] Suggest principal and subsidiary aims that might be involved in a choice of:

(a) type size for newspaper headlines;

(b) dimensions for microfilm or microfiche records of research articles;

(c) dimensions for a microdot;

(d) foreign characters for certain scientific or mathematical quantities (e.g. $\gamma$-rays, $\pi = 3.14159\ldots$).

# 2 · Symbols and Codes

## 2.1 BINARY ELEMENTS

All but a very few of the digital computers ever constructed have performed their internal operations using physical elements which could take just one of two states. In the earliest electro-mechanical machines the elements were relays and switches which could either make or break circuits. Later machines used electronic devices to achieve similar ends. Circuits are either passing current or they are not; areas of a magnetic surface are of one polarity or of the opposite. Each of these elements can, therefore, represent one piece of information of the 'Yes' or 'No' variety and if one has sufficient of them the number of distinct patterns of 'Yesses' and 'Noes' is great enough to be able to assign a unique pattern to each distinct item of information that it is desired to represent. There will be many ways of making this assignment, each way having its own properties.

The choice of the particular binary elements to be used in any computer has been governed primarily by considerations of simplicity, cheapness and reliability. The few cases where more complex elements were chosen to give a closer correspondence to the types of information to be represented did not have sufficient success to persuade machine designers to adopt them more frequently. For example, one machine which was intended primarily for arithmetic purposes used a device having ten states, presumably on the grounds that the greater ease of representing and manipulating numbers in the scale of ten would outweigh the greater cost of the devices themselves. There are still many examples of devices with more than two states in equipment rather less complex than computers, for example in digital clocks or those in which the second or minute hand makes discrete movements. For most practical purposes with computers we can restrict ourselves to considering representations of information by patterns of ones and zeros where the two states of the devices are themselves represented by the digits one and zero.

## 2.2 EXTERNAL REPRESENTATION

It has long been easy to recognize automatically whether a suitable piece of paper has a hole in it in a standard position or not. One of the first methods of recognition used metal pins which would pass through the holes but be stopped if no hole had been punched and those that passed through activated some mechanical or electrical trigger. Another method used an opaque paper so that light could pass through a hole and activate a photo-electric cell. The common media in use have been punched cards of various sizes and paper tape of various widths. Before some measure of standardization had been agreed, companies manufacturing computers and peripheral equipment took their own decisions about representation for the character sets to be used. The choices of representation usually incorporated some features which would give protection against errors - a sensible tacit admission that complete reliability could hardly be expected from a material as flimsy as paper and a recording method which involved some moving parts. The whole study of codes which could give some protection against errors became even more important when one had to transmit information over distances through some form of telephone network. Here, as elsewhere, the choice of representation involved compromises, some of which are illustrated in the following sections.

## 2.3 PAPER TAPE

A paper tape which has space for a sprocket hole near the centre of the tape and for $r$ information holes to be punched in its width can represent $2^r$ different symbols - since each position along the width can either have a hole punched or not. The nearly central sequence of sprocket holes serves to align the rows of holes over the reading stations; in early readers a small cog wheel engaged the sprocket holes and both aligned and drove the tape while in later ones a friction drive was substituted for the cog and the alignment done photo-electrically. Two of the characters almost force special consideration. For much equipment it is necessary to have some length of blank tape, that is tape with no holes punched at all, to act as a leader for threading the tape before reading. This in itself would suggest that a row of blank tape should not represent a character and the suggestion is reinforced by the advisability of guarding against the total failure of the punching mechanism. At the other extreme a row of tape with a hole punched in every position is conveniently reserved for the erase symbol. Clearly, whatever pattern of holes is punched in the tape, a superposition on it of the character causing all the holes to be punched will effectively erase whatever was there before. When these two patterns of holes are reserved for the blank and erase characters there remain $2^r - 2$ other possible combinations. The combinations fall into two sets, those in one set having an even number of holes punched and those in the other with an odd number of holes. The former are said to have <u>even parity</u> and the latter <u>odd parity</u>.

### 2.3.1 Parity

The simplest of the error detection criteria is based on checking the parity of the characters. If a set of symbols of the same kind, for example, all the decimal digits, are to be represented in a particular code by combinations with odd parity, a single error in one of the punch positions would yield a combination of even parity which could not represent a digit. If the context of the problem demands a digit then the error can be detected. The capability of detection by parity does not of course extend to errors simultaneously occurring in an even number of the positions on the tape. For example, an error in two positions will leave the parity of the combination unchanged and could turn one digit into another. But as such simultaneous occurrences of errors are much less likely than a single one, parity is a most useful criterion for detection especially when coupled with the special provisions for detecting blank and erase.

Fig. 2.1 shows a portion of the paper tape code used on an early machine. The characters important in numerical work are all represented by combinations of five holes with odd parity. One can notice too that the first column is used for the parity bit while the remainder of the

| Pattern | Figure shift | Character letter shift |
|---|---|---|
| 00001 | 1 | A |
| 00010 | 2 | B |
| 00100 | 4 | D |
| 00111 | 7 | G |
| 01000 | 8 | H |
| 01011 | - | K |
| 01101 | Line feed | M |
| 01110 | Space | N |
| 10000 | 0 | P |
| 10011 | 3 | S |
| 10101 | 5 | U |
| 10110 | 6 | V |
| 11001 | 9 | Y |
| 11010 | + | Z |
| 11100 | . | . |
| 11011 | Letter shift | |
| 00000 | Figure shift | |

Fig. 2.1. Pegasus/Mercury paper tape code

columns can be read as a binary integer to give the numerical value of the character being represented. Thus, the other columns carry weights of 1, 2, 4 and 8. As there are only 30 useable combinations in the five-channel code quoted, none can be left unassigned unless only a very small character set is to be represented. Indeed, if it is desired to represent the integers, the alphabet and a few other symbols, there needs to be a provision for interpreting any combination in one of a number of ways. For the particular code quoted, two combinations are reserved for shift characters which indicate that characters following them are to be interpreted as, for example, figures until the letter shift character is encountered. Thus the two different interpretations of one sequence of characters which are identical save for the first shift character could be 81 4.217 or HAND. BAG.

A tape code having eight hole positions across its width may, however, have sufficient combinations to make it necessary only to assign half of them to the set of characters to be represented. In this case, the parity of all the characters can be the same and thus a single error in a hole position detected by a check on parity alone. The code for the KDF9 computer was of this kind; all characters were represented by combinations with even parity and in fact the code used only six information bits for most of the characters, the other two bits being used, first for parity and the two characters erase and space.

## 2.4 ERROR CORRECTION AND DETECTION

Parity is one method of detecting an error in a code word. It is one way of dividing all possible combinations of a fixed number of binary digits - or whatever other symbols make up the code words - into two sets: one set containing acceptable code words and the other invalid combinations. Such a division of all the combinations of code symbols can be extended - the whole set can be divided into valid code words, the combinations arising from a given amount of corruption of each code word, and the remainder. If it can be arranged that the combinations in the second set could have arisen from an error in one symbol in one and only one valid code word or possibly from two or more errors in symbols from some other code words, then we may be prepared to decode such a combination as the equivalent to the unique valid code word from which it has suffered least distortion. In these circumstances such a division of the possible code combinations has led to a code which will correct a single error in a symbol as well as detecting it. If there are any other symbol patterns, all that can be said about them is that errors have been detected but they cannot be corrected under the simplest assumptions about how many such errors have occurred. This situation is illustrated in fig. 2.2: each correct code word has a region round it containing words differing from it by at most a given number of errors; these are the words which if received would be decoded in the same way as the corresponding correct word. The space between such regions corresponds to code words which can merely be flagged as in error. To place these concepts on a formal basis we need a concept of distance between two code words. A simple and appropriate function is the Hamming distance.

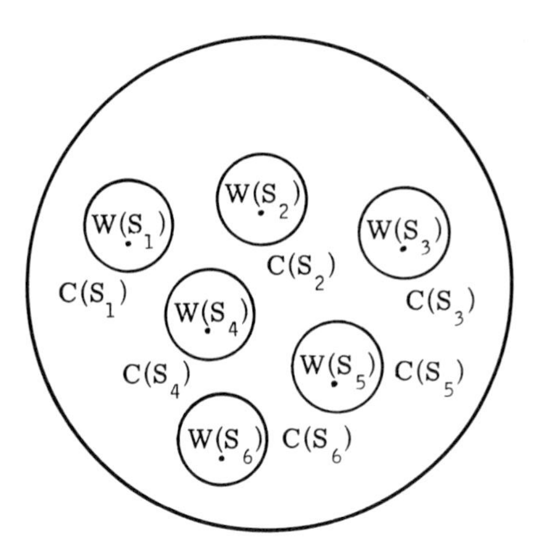


Fig. 2.2. Code words and correction regions

### 2.4.1 The Hamming Distances

**Definition.** Given two strings of symbols of the same length $W_1$, $W_2$ then the Hamming distance $d(W_1, W_2)$ is the number of positions in which the strings $W_1$ and $W_2$ differ.

Clearly this definition of a distance has the right sort of properties. The distance between any two strings is non-negative, symmetric, and the only distances between strings that are zero, are the distances between any given string and itself, i.e. $d(W_1, W_2) \geq 0$, $d(W_1, W_2) = d(W_2, W_1)$ and $d(W_1, W_2) = 0$ implies $W_1 \equiv W_2$. Further, the triangle inequality is satisfied

$$d(W_1, W_2) + d(W_2, W_3) \geq d(W_1, W_3)$$

so that the distance between two strings is no greater than total distance between those strings and any intermediate ones.

The condition for a code to be capable of detecting all cases of a given number of errors as well as all fewer errors can now be expressed in terms of the Hamming distance.

### 2.4.2 Detection of Errors

**Theorem I.** The necessary and sufficient condition for a code to be capable of detecting all cases of $k$ or fewer errors is that the Hamming distance between all pairs of code words is at least $k + 1$.

The proof is almost immediate; the condition is necessary because two code words that have a distance $r$ between them can be derived from each other by just $r$ errors in the symbols of the code word. Thus to satisfy the requirement of detecting all cases of $k$ or fewer errors we must have $r > k$. Conversely if the code words are all distant at least $k + 1$ from each other no occurrence of only $k$ errors can convert one code word into another. The condition is therefore sufficient.

This case of just detecting a number of errors is one in which the possible combinations of code symbols have been divided into two classes - the first class consists of the valid code words and the second of all the invalid combinations. If one wished also to be able to correct some errors in transmission as well as detect errors we need to divide all the possible combinations into more classes. Suppose that each message symbol $s$ has a corresponding unique code word $w(s)$ but has also corresponding to it a class of code symbol patterns $c(s)$. If a pattern which belongs to the class $c(s)$ is received then it is decoded as the message symbol $s$ in just the same way that the correct code word $w(s)$ is itself decoded. In order to avoid ambiguity we require that the two classes $c(s_1)$ and $c(s_2)$ have no members in common when $s_1$ and $s_2$ are different. There may (and in general will) be some patterns which do not belong to the classes $c(s_i)$ for any message symbol $s_i$: if such a pattern is received the most that can be said is that errors in transmission have been detected but that it is impossible to correct them. There is a limit to what can be achieved in detection and correction which is stated by theorem II.

### 2.4.3 Correction of Errors

**Theorem II.** The necessary and sufficient condition for a code to be capable of correcting all cases of $k$ or fewer errors is that the Hamming distance between all pairs of code words is not less than $2k + 1$.

If such an error correcting code existed in which a pair of code words were less than $2k + 1$ distant from one another, then there is certainly a pattern which could be regarded as having $k$ errors from one of these code symbols and $k$ or less errors from the other; accordingly the condition is necessary. Conversely the condition is sufficient; we can divide all the possible patterns of code words into classes $c(s_i)$, one class corresponding to each code word and the remainder $c$, say. The members of each $c(s_i)$ are chosen so that each is less than $k + 1$ from its corresponding message symbol $s_i$ and the $c(s_i)$ are clearly distinct. The remainder class, $c$, consists of those patterns which, when received, can merely signal the occurrence of more than $k$ errors but cannot indicate a correct code symbol.

Error correction and detection are capabilities which have some cost. In both cases we must sacrifice some of the possible patterns and not allow them to represent different message symbols. It follows that a set of $N$ message symbols will require more than $N$ code patterns to represent them with a given level of error detection or correction. Alternatively a given set of code patterns can be used to represent only a smaller set of message symbols under the same circumstances. Consider codes which are designed to correct a single error and which use code words of $n$ symbols each where the symbols themselves come from $b$ different characters. Then the number of message symbols that can be represented can at most be $N$ where

$$N \leq b^n/(nb - n + 1) . \qquad (2.1)$$

Since each of the possible patterns can have any one of the $b$ characters in each position, there is a total of $b^n$ patterns. For a given message symbol one of these patterns is its correct code representation and then there are $n(b - 1)$ other patterns in the class of those which are distant one from the correct code representation.

In the common binary case where the code symbols are just two characters, 0 and 1 say, we can consider how many binary digits must be added to each pattern of $m$ - bits representing the $2^m$ message symbols in order to allow a single error to be corrected. In this case the relation that places a lower limit on the number $r$ of extra bits is given by (2.1), with $N = 2^m$, $b = 2$, and $n = m + r$; hence we must have $m \leq 2^r - r - 1$. It has been shown that these bounds can be achieved for single error correction and codes have been designed for this purpose. It appears immediately however that the allocation of just one bit for checking ($r = 1$) is insufficient since the bound leads to $m = 0$. Even two bits can only be used to correct the single error in a set of just two message symbols which are coded by three bits. This of course follows immediately from theorem II which states that the Hamming distance between any two code words must be three; in the binary case with only three bits we are limited to just one pair of acceptable patterns and to using a 'two out of three' correct rule for the three bits.

## 2.5 CODING THEORY

Paper tape codes and the error-correcting codes so far considered have used code words all of which have the same number of code symbols. So far, also, we have only been concerned with coding in terms of sequences of binary digits. Coding theory studies generalizations of these concepts in a variety of ways. It is concerned with transforming the symbols of a message into code words which use the symbols of a code alphabet. It envisages a model which includes a source of messages which occur in some defined statistical manner, being received at an encoding device. This device produces a coded signal which is transmitted over an information channel which may, or may not, introduce errors. The received signal is passed to a decoder which aims to produce a message in the symbols of the original source alphabet. Coding theory is accordingly concerned with a description of the characteristics of the message source and with the construction and properties of codes and their suitability for use with various channels.

Codes so far encountered which use the same number of code symbols in each word are called block codes. Their use leads to particularly simple coding and decoding algorithms, e.g., in the reporting of the results of football matches, a win by the home team could be denoted by printing 1, a win for the away team by 2, a drawn game by X and an abandoned match by A. These four message symbols can be coded, using a code alphabet 0, 1 by the block code

$$1 \rightarrow 00, \quad 2 \rightarrow 01, \quad X \rightarrow 10, \quad A \rightarrow 11 .$$

Hence a sequence of results X112A1 ... can be transmitted in code symbols 100000011100... The decoding is unambiguous and simply consists of dividing the received string into pairs of code symbols and replacing the pairs by the message symbols. Other codes are possible, e.g.,

$$1 \rightarrow 0, \quad 2 \rightarrow 10, \quad X \rightarrow 110, \quad A \rightarrow 111 .$$

The earlier sequence of results would be represented by 11000101110... Such a string can again be decoded unambiguously and with only little more difficulty than the preceding one. For the results given, the number of code symbols required for the second code is one less than that when the first code is used. The number of code symbols used for the second code (and in general) depends upon the frequency with which the message symbols arise from the information source. Accordingly, the efficiency of a code as measured by the average number of code symbols required to represent the message symbols can only be calculated for arbitrary codes when the frequencies of the individual message symbols are known. If a message source produces symbols $s_1, s_2, \ldots s_n$ independently with probabilities $p_1, p_2, \ldots p_n$ respectively and are represented by code words with lengths $l_1, l_2, \ldots l_n$ code symbols, the average length of the code word is

$$L = p_1 l_1 + p_2 l_2 + \ldots + p_n l_n .$$

In our example, if the four results are equally likely, all the probabilities are 0.25 and the average length of coded messages for the second code is given by

$$\begin{aligned} L &= 0.25 \times 1 + 0.25 \times 2 + 0.25 \times 3 + 0.25 \times 3 \\ &= 2.25 \text{ code symbols per message symbol.} \end{aligned}$$

Equal probabilities of the four results is rather contrary to observation as teams playing at home seem to gain an advantage and matches are abandoned comparatively rarely. If the probabilities are

$$p(1) = 0.5, \quad p(2) = 0.25, \quad p(X) = 0.20, \quad p(A) = 0.05$$

the message source needs less code symbols on average for each message symbol

$$L = 0.5 \times 1 + 0.25 \times 2 + 0.20 \times 3 + 0.05 \times 3$$
$$= 1.75 \text{ code symbols per message symbol.}$$

In this case the average length of the second code is less than the length, 2, of the block code.

### 2.5.1 Information Units

In the general case we would like to discover the minimum number of code symbols required to code a population of message symbols which come from some source with known frequencies, whether it is possible to construct codes to achieve this optimum, and how to construct best possible codes. If less code symbols are needed on average to report the symbols produced from a message source, there is in some sense less information in the messages. In order to define a measure of the amount of information in a message, let an event $E$ have probability $P(E)$ of occurring. If we are told that $E$ has occurred, the amount of information that we have received is defined to be

$$I(E) = -\log P(E) \text{ units of information.}$$

If logarithms* are taken to base 2, the units are called 'bits'; thus if the event $E$ has equal probabilities of occurring and not occurring, $P(E) = \frac{1}{2}$, as in tossing an unbiased coin, the amount of information in reporting the result of such an experiment is one bit.

This definition of the amount of information has the desirable property that the amount of information in reporting the occurrence of two independent events, $E_1$ and $E_2$, is the sum of the amounts of information in reporting the occurrence of the two events separately. Since the events

* For the remainder of this chapter the base of any logarithms will be taken to be two.

are independent, we have $P(E_1 \text{ and } E_2) = P(E_1)P(E_2)$, hence

$$-\log P(E_1 \text{ and } E_2) = -\log P(E_1) - \log P(E_2) = I(E_1) + I(E_2) .$$

The definition also implies that we receive very little information when told that an almost certain event has occurred. The statement, 'The sun rose this morning', conveys no information in the terms of the definition. It is important to notice that information is not concerned in any way with the meaning of a message, but merely with the uncertainty of it occurring.

### 2.5.2 Entropy

We are now able to calculate the average information of a source S which emits message symbols in sequence, each symbol being independent of all others (such a source is said to have zero memory). Let the message alphabet be $s_1, s_2, \ldots s_n$ which occur independently with probabilities $P(s_1), P(s_2), \ldots P(s_n)$. Then the average information of

$$\begin{aligned}\text{the source} &= \sum_{i=1}^{n} P(s_i)I(s_i) \\ &= \sum_{i=1}^{n} -P(s_i)\log P(s_i) \\ &= H(S) .\end{aligned}$$

H(S) is called the entropy of the source and is on average the smallest number of binary digits that can be used to represent the symbols of the source. The source with the greatest entropy out of several with the same number of message symbols is that in which all symbols occur with equal probability, e.g. in the case of the football results the equiprobable source has entropy 2 bits per symbol. In the other case

$$\begin{aligned}H(S) &= -0.5 \log 0.5 - 0.25 \log 0.25 - 0.2 \log 0.2 - 0.05 \log 0.05 \\ &= 0.5 + 0.5 + 0.46 + 0.22 \\ &= 1.68 \text{ bits/symbol}\end{aligned}$$

$(\log_2 0.2 = \log_{10} 0.2/\log_{10} 2 = -0.6990/0.3010)$ .

Neither of the codes proposed in the example was able to achieve an average length as low as the entropy. Some reductions in length can be obtained by considering pairs, triplets, ... of message symbols and coding these combinations in an efficient way. A memory source consisting of all possible pairs of message symbols from a source s with their appropriate probabilities is called the second extension of the source S. In general the kth extension of a source with zero memory has entropy just k times that of the original source. Such a result does not hold for sources which do not have zero memory, e.g., the occurrences of pairs of letters in normal English text depart considerably from the frequencies that would be expected on a hypothesis of independence of the letters. Sources containing natural language strings have an amount of information much less than a calculation based upon the independence of the symbols would indicate.

## 2.6 CONSTRUCTION OF OPTIMAL CODES

If a code is such that all the code words are distinct it is said to be non-singular. If further the kth extension of the code is non-singular for every finite integer k, then it is also said to be uniquely decodable. If further a uniquely decodable code can be decoded without referring to any of the following symbols, it is instantaneous. These are all clearly desirable properties but some common codes do not possess them; the basic symbols of Algol are not instantaneously decodable, for example :, :=. We would like to be able to construct codes which are uniquely decodable and which have average length at most that of all other uniquely decodable codes for the same source S and using the same code symbols, i.e., to construct compact codes. An algorithm given by Huffman achieves this result. We consider here only codes using the two symbols 0, 1.

### 2.6.1 The Huffman Algorithm

Let the message symbols be $s_1, s_2, \ldots s_n$ arranged in order of decreasing probabilities so that $P(s_1) \geq P(s_2) \geq, \ldots P(s_n)$. Suppose that the corresponding code words of an optimum code are $W_1, W_2, \ldots W_n$ of lengths $l_1, l_2, \ldots l_n$ binary digits respectively. For an optimum code we must clearly have the message symbols of greater probability represented by the shorter code words since swapping the code words would increase the average length of the code, i.e.,

$$P(s_i) > P(s_j) \text{ implies } l_i \le l_j .$$

Further the two longest code words must be of the same length, for if not we could drop the additional digits in the longer word, i. e., $l_n = l_{n-1}$; thus code words of length $l_n$ occur in pairs which differ only in the last digit. The Huffman algorithm proceeds recursively. The two symbols of smallest probability have assigned to them different final digits for their code words and the pair are then replaced by another symbol which has probability equal to the sum of the two originals. The coding problem is thus reduced to that for a source with one symbol less than the original and the same procedure is repeated.

**Example.**

| $S_i$ | $P_i$ | $S_i$ | $P_i$ | $S_i$ | $P_i$ | $S_i$ | $P_i$ | $S_i$ | $P_i$ | $S_i$ | $P_i$ |
|---|---|---|---|---|---|---|---|---|---|---|---|
| A | 0.3 | A | 0.3 | A | 0.3 | A | 0.3 | A | 0.3 | B' | 0.4 |
| B | 0.2 | B | 0.2 | B | 0.2 | B | 0.2 | C' | 0.3 | A | 0.3 ]A' |
| C | 0.15 | C | 0.15 | C | 0.15 | D' | 0.2 | B | 0.2 ]B' | C' | 0.3 ] |
| D | 0.10 | D | 0.10 | F' | 0.15 | C | 0.15 ]C' | D' | 0.2 ] | | |
| E | 0.10 | E | 0.10 | D | 0.10 ]D' | F' | 0.15 ] | | | | |
| F | 0.05 | G' | 0.10 ]F' | E | 0.10 ] | | | | | | |
| G | 0.05 ]G' | F | 0.05 ] | | | | | | | | |
| H | 0.05 ] | | | | | | | | | | |

Hence we can code A' → 0, B' → 1; then A → 00, C' → 01;
C → 010, F' → 011; F → 0110, G' → 0111;
G → 01110, H → 01111.

Similarly B → 10, D' → 11; D → 110; E → 111.

This process can be represented graphically by the tree structure shown in Fig. 2. 3. The average length of this code is given by:

$$\begin{aligned} L(S) &= 2(0.3) + 2(0.2) + 3(0.15 + 0.1 + 0.1) + 4(0.05) + 5(0.05 + 0.05) \\ &= 2.75 \end{aligned}$$

which compares with an entropy of 2. 71. There are several places in this algorithm where one has a choice of position in the list for combinations of the symbols and each choice will lead to a different Huffman code for the source, all of them having the same average length. Some of the

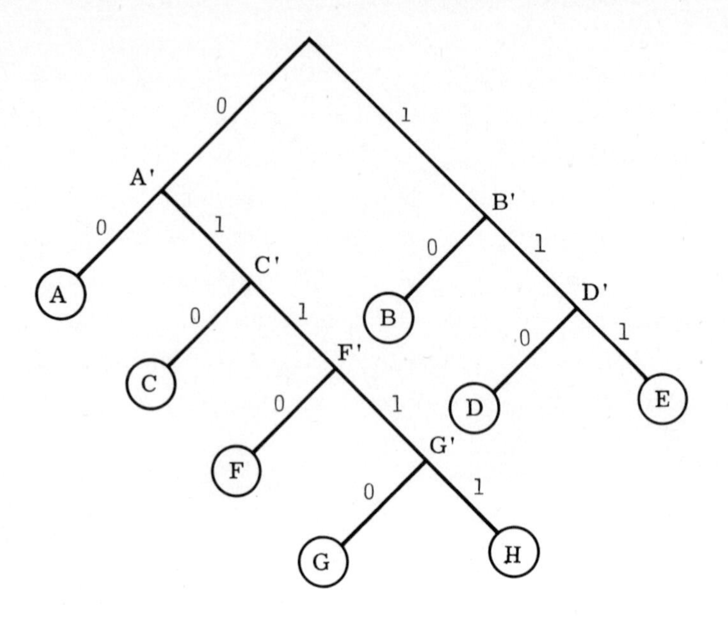


Fig. 2. 3. A Huffman coding tree

choices lead to codes in which the longest code words are as short as possible in any optimum code. These codes are achieved by inserting a combination of symbols like G and H as near the top of the list as one can without violating the descending order or probabilities. In our example the symbol G' would appear between C and D in the second column of symbols. If the example is worked through it will be found that G and H need only then be represented by four code characters although of course some other symbols have longer code words to compensate. The process corresponds to reducing the number of generations in the tree without violating the rules on combining symbols.

## 2. 7 WEIGHTED CODES

The example given in section 2. 3. 1 of a five-hole paper tape code represents the decimal digits by their binary equivalent in four of the hole positions and uses the fifth position solely to give odd parity. Accordingly, the digit represented can be obtained by adding together values assigned to the hole positions whenever a hole is punched there. The appropriate values for this code are 0, 8, 4, 2, 1 reading from left to right; thus 00111 represents $0 \times 0 + 8 \times 0 + 4 \times 1 + 2 \times 1 + 1 \times 1 = 7$. Such codes are known as weighted codes and several examples of them occur in computing applications. Just what is represented, and the weights, will be chosen to give some additional properties of the code, e.g. in the 'excess-3(XS3) representation' the decimal digits 0-9 are represented on four bits using the binary weights 8, 4, 2, 1 as above where the sum of the weights is 3 greater than the digit being represented (fig. 2.4).

| | | | |
|---|---|---|---|
| 0 | 0011 | 5 | 1000 |
| 1 | 0100 | 6 | 1001 |
| 2 | 0101 | 7 | 1010 |
| 3 | 0110 | 8 | 1011 |
| 4 | 0111 | 9 | 1100 |

Fig. 2.4. Excess three code

A fifth bit with weight 0 can of course be added for parity if required. This code has the property that the sum of two digits represented in it generates the carry of a binary digit in the fourth position, i.e. from the column with weight 8, when the two corresponding decimal digits would generate a carry into the tens position (see example 2.20).

A seven-bit weighted code which has been used in several machines is the biquinary. In this code the bits can be thought of as divided into two sets, one of two bits used only to indicate whether the digit is less than 5 or not and the other five bits just mark which decimal digit of the set of five is required by the position of a single 1. Accordingly, for the code shown in fig. 2.5 the weights are 5, 0, 4, 3, 2, 1, 0.

The device of weighting positions in a representation is one which will be encountered elsewhere in computing. In particular, it is the basis of a number of checking systems for numerical data intended for entry into computers.

| 0 | 01 | 00001 | 5 | 10 | 00001 |
|---|---|---|---|---|---|
| 1 | 01 | 00010 | 6 | 10 | 00010 |
| 2 | 01 | 00100 | 7 | 10 | 00100 |
| 3 | 01 | 01000 | 8 | 10 | 01000 |
| 4 | 01 | 10000 | 9 | 10 | 10000 |

Fig. 2. 5. Biquinary code

## 2. 8 PUNCHED CARD CODES

The history of punched card codes exhibits a variety of forms and sizes to rival the developments of paper tape and the codes used for it with, in this case, less easily discernable theoretical justification. Cards could hold 40, 60 or 80 columns of information, recorded by holes which were punched circular, rectangular, or diamond-shaped according to the manufacturer's equipment, in codes which the manufacturer had determined for his own purposes alone. The attempt to standardize on an 80-column card with a standard code using positions selected from 12 possible in each column, has been almost entirely successful and almost all modern computers either must have, or at least can have, reading and punching equipment for these 80 column cards (fig. 2. 6). Such standardization however is not quite complete as some small IBM computers now have their principal peripherals dealing with yet another size of card.

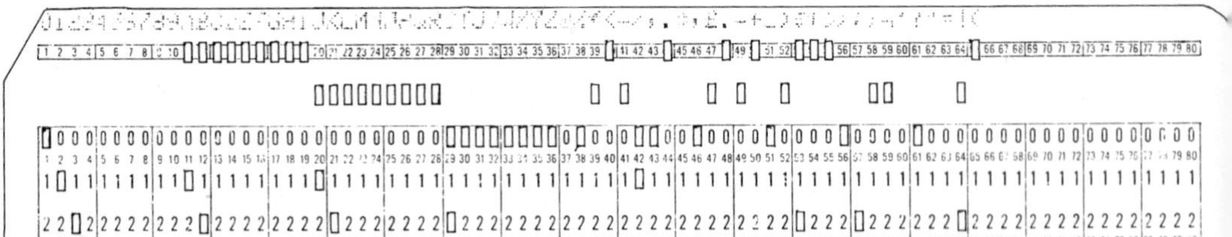

Fig. 2. 6. A punched card code

## 2.9 BIBLIOGRAPHY

The manuals of each of the early computers contained a description, even if not a justification, of the paper and punched card codes used. The current standard codes are described in the 'Standards' section of the Communications of the Association for Computing Machinery: see particularly vol. 13 (1970) p. 56 and p. 515.

There is a considerable literature on coding and information theory, requiring varying degrees of mathematical sophistication. The computing scientist can escape from needing a deep knowledge unless he gets heavily involved with communications or the detailed design of certain parts of computer hardware. One useful and readable book is Symbols, Signals and Noise by J. R. Pierce (Hutchinson, London); another, rather more advanced and mathematical is Information Theory by R. Ash (Interscience).

## EXAMPLES 2

[2.1] A tape punch for the code, part of which is shown in fig. 2.1, occasionally fails to advance the tape so that two characters are punched one on top of another. Will the parity check always detect this malfunctioning of the equipment? Will it never detect it? Can the erase symbol be produced by two decimal digits being punched consecutively? Give examples to justify your answer. How could errors of this kind be detected on input to a computer?

[2.2] What is the parity of

(a) 01100101,

(b) 10100100?

What is the Hamming distance between a and b? If one is given two binary strings of equal length, one of which has even parity and the other odd parity, is the Hamming distance between them necessarily an odd number?

[2.3] The four message symbols $w_1$, $w_2$, $w_3$, $w_4$ are coded

$w_1 \rightarrow 01100$
$w_2 \rightarrow 00011$
$w_3 \rightarrow 10000$
$w_4 \rightarrow 11111.$

(a) Write down the Hamming distance between all pairs of code words.

(b) Write down the classes of binary patterns which are a distance one from each of the code words. Do these classes have any members in common?

(c) How far is the string $s = 11001$ distant from each of the code words?

(d) What can be said about the error correcting and detecting capabilities of this code?

[2.4] The coding for a message source of five symbols is

$W_1 \rightarrow 0$
$W_2 \rightarrow 100$
$W_3 \rightarrow 101$
$W_4 \rightarrow 1101$
$W_5 \rightarrow 1110$

Can this code be optimal?

[2.5] A message source consists of eight symbols occurring independently with probabilities $\frac{1}{2}, \frac{1}{4}, \frac{1}{16}, \frac{1}{16}, \frac{1}{32}, \frac{1}{32}, \frac{1}{32}, \frac{1}{32}$. Construct a Huffman code and show that its average length is equal to the entropy of the source. What is the shortest block code that can represent the source?

[2.6] Construct a Huffman code for a source of 13 symbols with probabilities 0.20, 0.18, 0.10, 0.10, 0.10, 0.06, 0.06, 0.04, 0.04, 0.04, 0.04, 0.03, and 0.01. Calculate the average length of the code.

[2.7] Which of the following statements are true?:

The parity of a character in a paper tape code

(a) depends on the number of positions in which holes may be punched;

(b) depends on the number of holes which are punched;

(c) can be even or odd.

[2.8] A set of four characters $s_1$, $s_2$, $s_3$, $s_4$ are coded on five bits as follows:

$$s_1 \rightarrow 10110, \quad s_2 \rightarrow 11000, \quad s_3 \rightarrow 01101, \quad s_4 \rightarrow 00011 .$$

Which of the following are true?:

(a) This code uses one quarter of the possible five-bit patterns.

(b) The occurrence of errors in any two positions can always be detected.

(c) There is a Hamming distance of 2 between the codes for $s_2$ and $s_3$.

(d) The code alphabet used is (0, 1).

(e) If only a single error has occurred the pattern 01111 corresponds to $s_3$.

[2.9] If we want to encode two message symbols using a Hamming code to obtain single error correction, what is the shortest such code we can adopt? What is the corresponding answer for four message symbols?

[2.10] Define the concept of 'Hamming distance'. Describe how this concept is used in the design of error detecting and correcting codes.

Given that a discrete noisy memoryless channel carries binary digits in groups of five, design a code which will allow a single error in any group to be detected <u>and</u> corrected. How many distinct messages can be transmitted with this code?

(Liverpool 1969)

[2.11] What features are desirable in a character code for paper tape? In practice a compromise has to be reached. Illustrate this by reference to a code that you have met.

What is the maximum number of characters which can be represented on seven-track, odd parity paper tape:

(a) with one-to-one correspondence between character and code;

and (b) with a two case system.

What special significance is commonly attached to the all-ones symbol?

(Leeds 1969)

[2.12] A message source is coded

$$A \to 1111, \quad B \to 1100, \quad C \to 1010, \quad D \to 1001, \quad E \to 0110,$$
$$F \to 0101, \quad G \to 0011, \quad H \to 0000.$$

Can any errors be corrected? or detected?

[2.13] Check that the entropy of the source in the example in section 2.6 is $H(S) = 2.71$. Derive for this source a Huffman code with longest codes as short as possible and calculate its average length. Draw the corresponding tree.

[2.14] A zero memory source emits symbols $v_1, v_2, v_3$ each with probability $\frac{2}{11}$, and $c_1, c_2, c_3, c_4, c_5$ each with probability $\frac{1}{11}$.

What is the shortest block code which can code the source symbols? Derive two Huffman codes for the source, one with the longest code words as short as possible and the other with the shortest possible code words assigned to symbols with the highest probability, and compare their average lengths with that of the block code.

[2.15] Define the terms compact and instantaneous applied to a code for a source S. Derive a compact instantaneous binary code for the source which reports the number of heads in 10 independent tosses of an unbiased coin. Calculate the average length of your code and the entropy of the source.

(Newcastle 1969)

[2.16] An information source consists of message symbols which are all two digit octal numbers 00, 01, ... 77 which have equal probabilities.

What is the entropy of this source?

Suggest how the message symbols may be efficiently coded using the two symbols 0, 1.

How much information is given by the statements that a message symbol received from this source

(a) has two different digits,

(b) has both digits the same,

(c) is less than $40_{(8)}$,

(d) is odd?

(Newcastle 1970)

[2.17] A memory source, S, emits symbols $s_1$, $s_2$ with probabilities $\frac{3}{4}$, $\frac{1}{4}$ respectively. Construct the second extension of this source and a compact binary code for it. What is the average length of this code per source symbol of S? Compare it with the entropy of S.

[2.18] The 'genetic code' which is transmitted by DNA conveys its information through linear arrangements of four amino acids, but not all sequences represent distinct messages which, instead, are given by the fraction of the positions in the sequences occupied by the different acids. For the simple case of just two different acids A, B and sequences of three of them, compare the entropy of the sources for which all sequences are equally likely when (i) each represents a distinct symbol (ii) the distinct symbols are those represented by 0, 1, 2, 3 acids of type A in the sequence.

[2.19] Some mark readers will detect a pencil line drawn in a number of positions across a page. The amount of work encoding the symbols of a source is proportional to the average number of pencil marks that have to be made but no symbol may be represented by no mark at all. If marks may be made in three positions only and the source has seven symbols with probabilities 0.3, 0.2, 0.15, 0.15, 0.15, 0.03, 0.02, suggest an optimum code for the mark reader and calculate its average length.

[2.20] Find an algorithm for producing the tens and units digits in XS3 code of the sum of two decimal digits in that code.

[2.21] The decimal digits can be represented by a four-digit weighted code with weights 2, 4, 2, 1. Devise such a code which has the property that any two decimal digits which add to nine are represented by codes which have 0s in the position that the other has 1s (i. e. are 'ones-complements' of each other).

[2.22] Is the biquinary code (fig. 2.5),

(a) single error detecting

(b) single error correcting?

[2.23] A Gray code represents decimal digits, N, as four-bit words $g_0g_1g_2g_3$ according to the rules:

(1) express N as $x_0 2^0 + x_1 2^1 + x_2 2^2 + x_3 2^3$,

(2) put $g_3 = x_3$,

(3) for $i = 0, 1, 2$ put $g_i = x_i + x_{i+1} - 2x_i \cdot x_{i+1}$.

(a) Write down the code words for $N = 0, 1, 2, \ldots 9$.

(b) Is this code single error detecting?

(c) What is the Hamming distance between the words for N and $N + 1$?

[2.24] Define the terms instantaneous and uniquely decipherable with regard to codes.

Give an algorithm or testing procedure which will determine whether a code defined by the code words $x_1, x_2, \ldots x_n$ is uniquely decipherable. Show how your algorithm would operate on the following two codes.

Code A

| | | | |
|---|---|---|---|
| $x_1$ | 010 | $x_5$ | 00011 |
| $x_2$ | 0001 | $x_6$ | 00110 |
| $x_3$ | 0110 | $x_7$ | 11110 |
| $x_4$ | 1100 | $x_8$ | 101011 |

Code B

| | | | |
|---|---|---|---|
| $x_1$ | abc | $x_5$ | bace |
| $x_2$ | abcd | $x_6$ | ceab |
| $x_3$ | e | $x_7$ | eabd |
| $x_4$ | dba | | |

Construct an ambiguous sequence for any code which you find to be not uniquely decipherable.

(Newcastle 1974)

[2.25] A zero memory source emits 155 different symbols and these symbols can be divided up into the following groups:

Group A - the five most frequent symbols which each have probability 0.1.

Group B - the next ten most frequent symbols which have probability 0.02.

Group C - the next twenty most frequent symbols which have probability 0.005.

Group D - the next thirty most frequent symbols which each have probability 0.003.

Group E - the next forty most frequent symbols which each have probability 0.0015.

Group F - the last fifty symbols which each have probability 0.001.

(a) Construct a Huffman code for the groups (A-F).

(b) What is the length of the shortest block code for the symbols themselves?

(c) If you are restricted to 4-bit codes or multiples of 4 bits how could you construct a code for the symbols using just two escape characters?

(d) Could you improve on the code constructed in part (c) if you are allowed any number of escape characters? If so give the code.

In each case give the average length of the code.

(Whenever a code is mentioned in this question a binary code is required.)

(Newcastle 1976)

# 3 · Internal Representation

## 3.1 UNITS OF STORAGE

Since internal storage in computers is composed of elements which can take two states, the size of a store can be described by the number of bits that it contains. In these terms the store of a basic mini-computer would contain a few tens of thousands of bits and that of a large system some tens of millions. Of course the items of information that computers manipulate are normally represented by patterns of several binary digits and it is convenient to have some larger units of storage. Since the introduction of the IBM system 360 range of computers a commonly used unit has been the byte which is composed of eight bits. There are $2^8 = 256$ different states that can be represented by a byte and this is ample for the representation of all the letters, digits and symbols on a normal keyboard. Not all machines, nor even all machines from the same manufacturer are organized to manipulate bytes. Some use sets of six bits called characters (or even, confusingly, bytes) with or without a specification of how many bits are in them.

Both bytes and characters are convenient small units of storage but something still bigger is needed for many of the internal operations in the machines. Although most modern computers of medium size and greater have instructions for manipulating single bytes their main repertoire of instructions deals with larger sets of bits called words. Typical lengths of words in medium and large computers are thirty-two or twenty-four bits while mini-computers use words of twelve or sixteen bits but earlier machines have displayed a variety of word lengths. When store sizes are expressed as a number of words, comparisons between different machines need particularly to take account of the length of the words. Naturally, how much store is available for any given application depends also on other factors like the types of instruction available and their format, as well as system software.

On some modern machines the main memory is also regarded as being divided into 'pages' consisting of a number of contiguous bytes of storage. A typical size is 4096 bytes. Such a page is a convenient unit for transfer between main memory and backing storage devices, particularly in multi-programming and time-sharing systems, although arguments have been advanced for other sizes of page.

The contents of a word, as stored in the computer can be recorded in a variety of ways. The most straightforward is to use the binary integer representing the pattern of bits in the word; fig. 3.1 shows a 24-bit word.

| | |
|---|---|
| Binary | 0 1 1 1 1 0 1 0 1 1 1 1 0 0 1 0 1 0 0 1 1 0 1 1 |
| Octal | 3 6 5 7 1 2 3 3 |
| Hexadecimal | 7 A F 2 9 B |

Fig. 3.1. Recording a 24-bit word

The octal representation obtained by giving the numerical values of consecutive sets of three bits is shown below; the octal number requires only eight symbols instead of the twenty-four of the binary representation, and nothing other than the digits normally available on any keyboard is needed. Octal is convenient for producing core dumps in machines organised in six-bit byte units. Those using machines with bytes of the more normal eight bits are likely to encounter the even more compact hexadecimal representation shown in the last line of the figure; each byte is made up of two hexadecimal digits denoted by the characters 0, 1, ... 9, A, B, C, D, E, F which are equivalent to the decimal values 0, 1, ... 9, 10, 11, 12, 13, 14, 15 respectively.

## 3.2 CONVERSION BETWEEN SCALES

In the 'old' days of computers (the nineteen-fifties and even a little later) it was necessary to be able to read displays and printing in binary format. More recently most software designers have made provision for printing diagnostic and other information in the form that the users want but even now there will be occasions when some hand conversion between scales is unavoidable.

### 3.2.1 Decimal Integers to Binary

If the decimal integer is repeatedly divided by 2, the remainders give the binary digits, least significant first. For example,

$$
\begin{aligned}
7512 \div 2 &= 3756 \quad (+\ 0 \text{ remainder}) \\
3756 \div 2 &= 1878 \quad (+\ 0) \\
1878 \div 2 &= 939 \quad (+\ 0) \\
939 \div 2 &= 469 \quad (+\ 1) \\
469 \div 2 &= 234 \quad (+\ 1) \\
&\text{etc.}
\end{aligned}
$$

Thus $(7512)_{10} = (111\ 010\ 1011\ 000)_2$.

The same result is obtained by repeatedly subtracting the highest power of 2 possible which generates the binary integer from the most significant end.

Thus

$$
\begin{aligned}
7512 - 4096 \text{ (i.e. } 2^{12}) &= 3416 \\
3416 - 2048\ (2^{11}) &= 1368 \\
1368 - 1024\ (2^{10}) &= 344 \\
344 - 256\ (2^{8}) &= 88 \\
&\text{etc.}
\end{aligned}
$$

### 3.2.2 Decimal Integers to Octal

The routine of repeated division by the base of the scale is clearly applicable generally. Conversion of $(7512)_{10}$ to octal would proceed as follows:

$$
\begin{aligned}
7512 \div 8 &= 939 \quad (+\ 0 \text{ remainder}) \\
939 \div 8 &= 117 \quad (+\ 3) \\
177 \div 8 &= 14 \quad (+\ 5) \\
14 \div 8 &= 1 \quad (+\ 6)
\end{aligned}
$$

Hence $(7512)_{10} = (16530)_8$.

Notice that rapid conversion can be effected between the binary scale and octal (or any power of 2 scale); the binary integer is first partitioned into sets of three bits starting from the binary point and converted set by set into the octal integers.

Thus

$$
\begin{aligned}
(7512)_{10} &= (111\ 010\ 1011\ 000)_2 \\
&= (1|110|101|011|000)_2 \\
&= (1\ \ 6\ \ 5\ \ 3\ \ 0)_8.
\end{aligned}
$$

Similarly $\quad = (1\ D\ 5\ 8)_{16}$ where $(D)_{16} = (13)_{10}$.

### 3.2.3 Decimal Fractions to Binary

Since a binary fraction like 0.1011 represents $1 \times 2^{-1} + 0 \times 2^{-2} + 1 \times 2^{-3} + 1 \times 2^{-4}$, a decimal fraction may be converted into binary by repeatedly doubling it, subtracting and noting any whole number parts that occur.

For 0.6875 this yields (1).375, (0).75, (1).5, (1).0, so that $(0.6875)_{10} = (0.1011)_2$.

The doubling procedure for a conversion of binary fractions generalizes easily to other scales: for octal, one multiplies repeatedly by 8, hexadecimal by 16 and so on.

E.g. $\quad (0.6875)_{10} = (0.54)_8$.

The conversion between bases which are powers of two can be achieved via the binary scale for fractions just as it was for integers by considering groups of bits starting at the binary point.

E.g. $\quad$ Since $(5)_8 = (101)_2$, $(4)_8 = (100)_2$

$$(0.54)_8 = (.101100)_2 = (.1011\ 0000)_2 = (.B0)_{16}.$$

If one has to convert anything other than short numbers (integers or fractions) from decimal into binary by pencil and paper alone, a good deal of writing can be avoided by converting first to octal and then replacing the octal digits by the corresponding three binary digits.

### 3.2.4 Conversion into Decimal Fractions

An integer $\alpha_1 \alpha_2 \ldots \alpha_n$ in the scale of b can be conveniently evaluated in decimal by a nested multiplication

$$(\ldots(((\alpha_1 b + \alpha_2)b + \alpha_3)b + \alpha_4) \ldots + \alpha_n).$$

For example

$$\begin{aligned}(16530)_8 &= ((((1 \times 8 + 6)8 + 5)8 + 3)8 + 0) \\ &= (((14 \times 8 + 5)8 + 3)8 + 0)\end{aligned}$$

and so on.

Fractions can be converted in a similar way, by nested division starting from the least significant end. The fraction

$$0.C_1C_2 \ldots C_{n-1}C_n \text{ converts into } (C_1 \ldots + \frac{1}{b}(C_{n-1} + \frac{C_n}{b}))).$$

For example

$$(0.1011)_2 = \tfrac{1}{2}(1 + \tfrac{1}{2}(0 + \tfrac{1}{2}(1 + \tfrac{1}{2})))_{10}$$
$$= \tfrac{1}{2}(1 + \tfrac{1}{2}(0 + \tfrac{1}{2}(1.5)))_{10}$$

etc.

## 3.3 INTEGERS: PACKED DECIMAL

One method of representing integers, encountered mainly in commercial applications, is in packed decimal format; in this, one decimal digit in the range 0-9 is stored in each of the two hexadecimal positions forming one byte. The advantage of such a representation is that there is no need for calculations to convert from one scale to another; it is thus quite a suitable choice when most of the operations on the data concern transfers between the various stores and between the input and output devices. Arithmetic operations are rather less convenient and the packed decimal representation is only adopted when there is little arithmetic to be performed on the data. The representation is wasteful of storage for all but very small integers - in packed decimal form one byte can store an integer only in the range 0-99, while eight bits using binary can range from 0 to 255.

In representations of integers in this form the sign of a number is indicated by one of two characters in one of the half-byte positions, giving an example of a 'sign and modulus' representation. The pattern used to represent the two signs + and - differ in two of the most widely used codes - the Extended Binary Coded Decimal Inter-Change code (EBCDIC) and the American Standard Code for Information Interchange (ASCII). In some machines the half-byte containing the sign is stored at the least significant (right-hand) end of the binary coded decimal number. The machine instructions for manipulating these numbers, of course, take account of the position of the sign.

## 3.4 INTEGERS: BINARY REPRESENTATION

Representation of numbers by their magnitude and their sign presents some slight complications in the elementary operations needed to add two numbers of different signs in addition to those inherent in the decimal representation. Consequently an entirely different method of representing negative numbers is adopted together with a true binary notation when there is any appreciable amount of arithmetic to be performed on numbers. In a binary representation the individual bytes in a word do not correspond to one or more of the decimal digits; instead the whole word (or in some machines, a half word) is used to specify positive integers by their true binary value together with one bit at the most significant end of the number to show the sign. Negative integers are represented by their complementary form. The most common form is the twos complement in which the integer $-x$ is represented by the positive integer $2^n - x$ in the full $n$ bits available; thus the most significant bit of the word is a sign digit taking the value 0 for positive numbers and 1 for negative ones. If the binary digits of the representations are $x_n \ldots x_1$, the integer value is given by

$$X = -2^{n-1}x_n + \sum_{i=1}^{n-1} 2^{i-1}x_i .$$

The range of integers that can be represented with a word of $n$ bits is $-2^{n-1}$ to $2^{n-1} - 1$.

Arithmetic operations are easy to perform on numbers in the twos complement representation. For example, a number is negated by replacing 0s by 1s and vice versa and then adding 1 in the least significant position. An examination of the different cases of addition and subtraction of integers shows that where results lie within the range of numbers that can be represented they are obtained correctly and moreover, that when the arithmetic result is out of range a simple test is available to detect it.

An alternative form of representation for negative numbers which has been adopted for some machines is the 'Ones complement'. In this form $-x$ is represented as $(2^n - 1) - x$ so that the sign bit has an apparent value $(2^n - 1)$. Numbers are easily reversed in sign in this representation - the bits themselves are just complemented; and zero has two forms, all noughts or all ones. The mechanism for addition needs to include provision for feeding any carry digit created from the most significant end of the numbers round to the least significant place and adding it - 'end-around carry'. An examination of the eight possible combinations of the most significant digits of the two numbers being added and the carry digit into that position show that the 'end-around carry' procedure produces the correct sum except where the sum has overflowed.

## 3.5 FRACTIONS: FIXED BINARY

The binary representation of integers which has just been described can serve to represent fractional numbers in the range -1 to 1 with little change. All that is needed is to regard all the integers as being multiplied by the factor $2^{-n+1}$. In the twos complement representation of fractions, the largest and smallest quantities that can be stored are respectively $1 - 2^{-n+1}$ and -1. Thus when fractions are so represented, the operations are performed as though there were a binary point immediately following the most significant binary digit which is the sign digit; this contrasts with the integer representation where the position of the imaginary point is to the right of the least significant digit.

In early computers the fixed point representations of numbers were the only ones available and any programs performing arithmetic had to take care to adjust the numbers in the computation by scale factors to keep them within the permitted range; the scale factors themselves had to be adjusted from time to time to preserve accuracy as well. The programmer was thus forced to be aware of the dangers of losing significant figures in arithmetic operations as well as being reminded by the occurrence of an 'overflow' failure of unexpected fluctuations in the size of numbers in the calculations. Fixed point representation thus had mixed blessings.

The casual and non-specialist users of computers found the complexity of performing arithmetic operations in general calculations difficult or tedious; they welcomed a representation of numbers which eliminated most of the problems of this range - the 'floating point' form.

## 3.6 FLOATING POINT NUMBERS

In the number representations so far considered, the decimal or binary point has been regarded as fixed at one end or other of the word. The restrictions on the range of numbers that can be so represented has led to an alternative one being adopted for most extensive calculations, in particular, for arithmetic operations on numbers of type **real** (in the Algol sense). For this purpose a number is represented in the computer in 'floating point' form by a pair of numbers $(x, y)$ whose value is taken to be $x.k^{y-c}$ where $k$ and $c$ are fixed integers. In several early machines there were no special hardware operations for performing arithmetic and manipulative operations on the pairs $(x, y)$; even some modern machines lack such operations by the choice of the designer on economic grounds - for example, the infrequency of needing them in commercial calculations or in certain minicomputer applications. The floating point operations can, of course, be constructed using a set of general purpose fixed point machine orders but more powerful ones are usually provided by hardware in all but the smallest of present-day general purpose machines; in the latter case the machine designer will have had to decide how to represent both $x$ and $y$ and to what accuracy, and to choose the integer values of $k$, called the base of the floating point representation, and $c$, the excess.

The most common choices of base are $k = 2$ (floating binary) or for byte oriented machines like the IBM 360/370 series $k = 16$ (floating hexadecimal). Usually the pair of numbers $(x, y)$ is packed into a single or a double word with $x$ as a signed fraction in some of the bits and $y$ as an integer in the others. For example, the designer of a binary machine having chosen a word length of 48 bits, might assign 40 bits for the value of $x$ and the sign of the number and the other 8 bits for the integer $y$ for the exponent of the base $k = 2$. The eight bits for $y$ are enough to represent integers from 0 to 255 inclusive which leads to 128 as a suitable value for the excess $c$ in order to give the range of the exponent from -128 to +127. All of the options for representing fractions are open to the designer but the two most commonly encountered are twos-complement and sign-and-modulus; in either case the value of x is usually normalised by a hardware operation to lie between $\frac{1}{2}$ and 1 (between $\frac{1}{k}$ and 1 in the general case) so that no space in the word is wasted with leading zeros. The number of bits assigned to the exponent (or characteristic) y governs the range of numbers that can be represented, while the bits for x determine the precision of the representation. In the previous example with 39 bits for x when normalised and one bit for its sign the precision is about $39 \log_{10} 2 \doteqdot 11.7$ decimal digits.

Some mini-computer designers, constrained to quite short word-lengths have adopted more complicated hardware solutions in order to squeeze a little more accuracy; since the most significant bit of a normalised binary fraction x must be '1', there is clearly no need to store it, provided that the arithmetic operations are properly designed to deal with numbers in this form - the extra expense of the more complicated circuitry may be an acceptable price to pay in order to obtain one extra bit of precision.

The floating hexadecimal representation in a single word of 32 bits in a machine of the System 360/370 type is by sign and modulus for the fractional (x) part; the sign is in the extreme left bit and $|x|$ is contained in binary in the rightmost three bytes (i.e. 24 bits). The exponent occupies the other 7 bits of the first byte and, since $2^7 = 128$, the excess c is conveniently taken as 64. Numbers are normalised so that the most significant hexadecimal digit of the fraction, x, is non-zero;

$$\text{i.e. } \frac{1}{16} \le x \le 1 - \frac{1}{16^6} .$$

The layout of the representation is shown in Fig. 3.2. Since it is possible

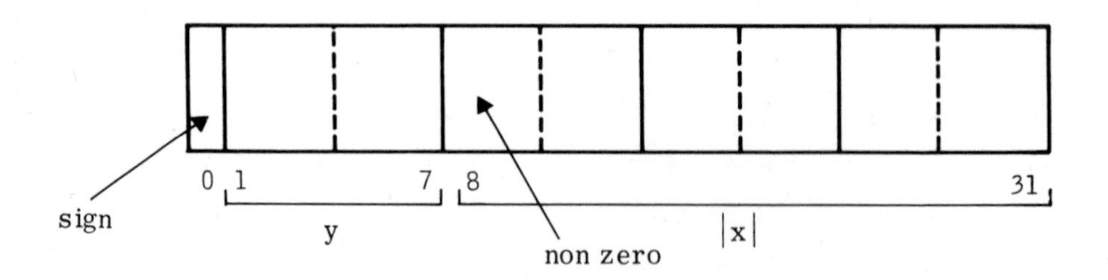


Fig. 3.2. Normalised floating hexadecimal

for bits 8-10 to be zero, x can be represented by only 21 bits and thus have a precision of between 6 and 7 decimal digits.

Although working with floating point numbers apparently relieves the programmer of most of the worry about the size of numbers and the scaling that fixed point representations require, loss of accuracy is still a danger and the warning signs are less obvious. The subtraction of two nearly equal numbers represented in floating point form loses significant figures just as it does for a fixed point representation: the operation of subtraction (or, of course, addition) of two floating point numbers $(x_1, y_1)$, $(x_2, y_2)$ first makes the exponent of the smaller the same as that of the larger and shifts the fraction to compensate, performs the subtraction (addition) on the two fractions and then normalises, adjusting fraction and exponent appropriately. Two nearly equal numbers have exponents the same or differing by at most one; after the exponents are adjusted the fractions are almost equal and their subtraction yields a number with several leading zeros and correspondingly fewer significant figures; the normalisation cannot restore those lost significant figures although the operation introduces new figures (usually zeros) at the less significant end of the fraction in order to fill out the bits which store it.

For example, let the two numbers to be subtracted be

a = 4 0 1 0 0 0 0 0

and b = 3 F F F F F F 0

in the usual 360/370 floating hexadecimal form. Then b is the smaller and is adjusted to

b' = 4 0 0 F F F F F.

The subtraction gives the un-normalised result having only one significant bit

4 0 0 0 0 0 0 1,

which normalises to

3 B 1 0 0 0 0 0.

Of course, if a or b has less than the maximum number of bits significant their difference has none at all.

While the catastrophic situation described above may be rare, there is usually a progressive, even if gradual, loss of precision from cancellations and rounding errors in a lengthy calculation and the number of decimal digits of precision provided in a single word has not been sufficient for many purposes. Accordingly provision is usually made for a double word representation and for operations on them as well. Another whole word may be assigned to the fraction as in the 360/370 machines, so that bits 8 to 63 (14 hexadecimal digits) are available giving about 16 decimal digits of precision. Alternatively, the second word itself can be formed as a valid floating point number with its exponent adjusted to give the magnitude of this portion of the number represented by the double length mantissa and the exponent in the more significant part of the double word. Even this extra precision may not be enough and another pair of words in a similar format can be assigned (cf. Extended precision working on 370 machines).

It is worth noting that the value 0 receives special treatment in many floating point representations. The most usual representation is to ensure that all the bits are zero so that this particular value is the same in both fixed and floating point representations. There is also usually provision for replacing by zero, numbers arising in the course of a computation (by multiplying two small numbers, usually) which are less than the standard representation will allow. This is the situation known as exponent underflow and will set a marker in some machines.

## 3.7 BIBLIOGRAPHY

The representation of numbers in the various forms permitted will be described in the manual for each computer, and often some detailed examples of the operations available on such numbers are given. More general accounts are presented in most texts, for example in Knuth vol. 2, and in Automatic Data Processing by Brooks and Iverson.

## EXAMPLES 3

[3.1] Express the decimal numbers 92 741, 2385, 0.2743, 0.0865 in binary.

[3.2] Express the binary numbers 1101011001, 10001110, 0.1101011001, 0.00100111 in octal, hexadecimal and decimal.

[3.3] Convert the binary number

111.111

to decimal form and the decimal number

111.111

to binary form, to a suitable number of binary places.

(Sheffield 1970)

[3.4] Explain the meaning of the terms

(i) sign and modulus,

(ii) twos complement,

(iii) ones complement,

with regard to the representation of integers in terms of binary digits. Show how the twos complement representation allows an adder designed for use with positive numbers to operate on quantities of either sign without modification.

(Leeds 1970)

[3.5] Explain why it is customary to manufacture computing systems from assemblies capable of taking up one or the other of two states. Give three examples of such devices.

Explain how to use such two state devices to represent

(a) English text,

(b) signed integers.

How can arithmetic operations be performed on numbers represented in the way that you have described?

(Leeds M.Sc. 1971)

[3.6] Write down algebraic expressions for the value of the N-bit signed binary integer

$$a_{N-1}a_{N-2} \dots a_2 a_1 a_0$$

appropriate to each of the twos complement, ones complement and sign and modulus methods of representation.

Hence or otherwise prove that, in twos complement notation, the representation of a signed binary integer in N + 1 bits may be derived from its representation in N bits by repeating the left-most bit.

What are the corresponding results in the ones complement and sign and modulus notations?

(Essex 1971)

[3.7] What are the answers, for each system of representation, to the following integer sums?

010010 + 011011
100011 + 100101
110100 + 111000
010111 - 011000

(Manchester 1971)

[3.8] Give an example of how a 48-bit word may be used to represent numbers in floating point binary. Compare the set of numbers which can be represented exactly in this way with the set available using twos complement representation of integers. How many numbers are there in each set?

Draw flow diagrams for algorithms which respectively add and multiply floating point numbers represented as you describe. You can assume the availability of fixed point addition, subtraction, multiplication and shifting operations.

(Leeds 1971)

[3.9] A floating point number in a certain computer is to be represented by 16 bits: one for the sign, five for the characteristic part and ten for the mantissa.

Give details of one such representation. What are the greatest and least positive, and greatest and least negative numbers permitted in your representation? How is zero represented?

With this representation in mind, what criteria would you use to define a numerical solution of a formal mathematical equation, $g(x) = 0$? State clearly any assumptions you make about the nature of $g(x)$.

(Newcastle 1968)

[3.10] Derive methods for converting a decimal number to a positional representation with radix $q$ and for performing the reverse transformation. Using these methods, or otherwise, convert the decimal number $107.581_{10}$ to octal notation and the octal number $263.417_8$ to decimal notation correct to six figures.

(Belfast 1969)

[3.11] Explain how a floating point number may conveniently be represented in a word-organized binary computer.

In the absence of hardware for direct arithmetic operations on such numbers, what operations are needed either for the addition or for the multiplication of two such numbers, leaving the result in the same form as the operands? You may assume the availability of logical, shift, and fixed point arithmetic instructions operating on 1-word operands, and of a double length accumulator.

If you were preparing routines, for general use, to carry out all four arithmetic operations on such floating point numbers would you provide these as macro-instructions, as subroutines or as an interpretive scheme? Give reasons for your choice.

(Leeds 1969)

[3.12] In what ways is it less efficient to represent integers as 'packed decimals' rather than as binary integers?

A new machine is to be based on a three-state device. Integers within this machine will be represented as

$$x = \sum_i t_i 3^i$$

where each $t_i = -1, 0,$ or $+1$, denoted by $\bar{1}$ 0 1 respectively.

(a) Write down the representations of 0, +17, -31 in this system using four-place numbers.

(b) How can numbers in this system be negated?

(c) Approximately how many places would be required to represent numbers as large as $\sim 10^{40}$, to an accuracy of about 12 decimal places?

(Leeds M. Sc. 1969)

[3.13] The specification of a certain compiler for Fortran on the IBM 360/67 computer implies that single-length real numbers are represented in floating point hexadecimal form using four bytes (32 bits). One byte contains the sign and the characteristic (k) plus sixty-four, where $-64 \leq k \leq 63$. The other three bytes together carry six hexadecimal digits representing the mantissa (m), where $\frac{1}{16} \leq m < 1$. Thus positive numbers can be represented approximately in the range

$$16^{-65} < 16^{k} \times m < 16^{+63} .$$

The following is a Fortran program:

```
   REAL A, B, C
30 FORMAT (F12.5)
   A = 722.0
   B = 0.85644531
   C = A + B
   WRITE (6, 30) C
   CALL EXIT
   END
```

The result output was 722.85620, but when the assignment to B was changed to:

B = 0.85644532

the new result was 722.85645.

Suggest an explanation of these two apparently inconsistent results. (For those not familiar with Fortran, the WRITE statement outputs C to device 6 in the format indicated at label 30, i. e. one number with twelve print positions and five digits after the decimal point. )

(Newcastle 1970)

[3.14] In a computer, signed fixed point integers are represented in twos complement form by the N bits $a_N a_{N-1} \ldots a_1$. Give an algebraic expression for the value of this integer. What range of integers can be represented? Calculate the value of the ten bit signed integer 1101011001.

Show that:

(i) the representation in this form of a signed integer in N + 1 bits may be derived from its representation in N bits by repeating the most significant bit; and

(ii) that an adder designed for use with positive numbers needs no modification to deal with negative numbers.

(Newcastle 1973)

[3.15] Describe the format of a hexadecimal floating point number in a single length (32 bit) register of a 360/370 type machine, and state the condition that must be satisfied if the number can be said to be normalised.

(1) 5 5 4 A 0 0 0 0

is a hexadecimal pattern in this format. What is its value? Normalise the pattern

(2) 5 5 0 8 0 0 0 0

Describe the operations that the floating point hardware unit has to perform in order to add together two floating point hexadecimal numbers and illustrate them by showing the stages necessary to add (1) to the normalised equivalent of (2), giving the resulting pattern.

(Newcastle 1974)

[3.16] Give the twos complement fixed point binary representations in 16 bits of -320, -0.75.

The designer of a computer proposes to represent floating point numbers to base 2 in a 32-bit word $x_1 x_2 \ldots x_{32}$ $(x_i = 0, 1)$ with an exponent excess 64 in bits $x_2 \ldots x_8$ and the fraction in twos-complement form in bits $x_1 x_9 x_{10} \ldots x_{32}$.

About how many significant decimal digits can be represented?

Represent 320, -320, 0.75, -0.75, 320.75, -320.75 in this form and give an algebraic expression for the value of the general binary pattern $x_1 x_2 \ldots x_{32}$.

How does the floating point representation in a word in the IBM 360 type machines differ from the form above?

(Newcastle 1975)

[3.17] Mention two of the main considerations influencing a computer designer in the choice of representations for floating point numbers. Give the form of floating point numbers in IBM 360 type machines. State precisely (a) the range of numbers that can be represented and (b) how many different numbers can be represented exactly.

A group of 360/370 computer users has a large part of its computations with complex numbers $z = a + ib$, and the group is considering writing a suite of programs to manipulate numbers of this form. They propose representing the pair (a, b) in the 'block' floating point form by (A, B, C) where C is an integer used as a hexadecimal exponent, and $a = A.16^{C-k}$, $b = B.16^{C-k}$, where k is a suitably chosen excess and the numerically larger of A, B is normalised. What arguments are there to advance for and against this proposal?

(Newcastle 1976)

[3.18] Explain the meaning of the terms

(a) twos complement,

(b) ones complement,

(c) sign and modulus,

with regard to the representation of integers in terms of binary digits. How would you interpret the 10-bit signed integer 1011101101 in each of the above three systems?

Integers are held in a three state machine in the following representation

$$x = \sum_i a_i 3^i$$

where $a_i = -1, 0, 1$ are denoted $\bar{1}, 0, 1$ respectively (e.g. $8 = 10\bar{1}$). The representation is known as the 'balanced ternary notation' and has base 3. In the balanced ternary notation

(i) Write down the representations of the decimal integers $5, -42, -10^4$, and 20.

(ii) How can we simply recognise the sign of a number?

(iii) Approximately how many places would be required to represent numbers as large as $\sim 10^{70}$ to an accuracy of 10 decimal places? (Take $\log_{10} 3 = 0.48$.)

(Newcastle 1975)

# 4 · Information Structures 1: Arrays

## 4.1 INTRODUCTION

In the last chapter some methods of representing single items of information usually within just one computer word were discussed; in the next three chapters we consider the representation of collections of items of information which can be related to each other in some way; for example, the co-ordinates of a point, the names of a father and his children or the location of a house and its number and type of rooms. Notice that the items of information may, but need not, be all of the same types - all numbers or all letters for example. Such collections of information are called information (or data) structures.

Simple examples of information structures are arrays as used in Fortran, Algol, and PL/I and structures as in Cobol and PL/I. In this first chapter on information structures we will examine arrays which are the most commonly used and simplest type of information structure. In the next chapter we will examine linear lists of elements, which in fact include arrays as a special case. In particular we will examine the difference between sequential storage and linked storage of such lists. In the third chapter on information structures we will extend these ideas to trees, which are the most important non-linear structures; they have hierarchical and branching relationships between the elements. In all cases we will principally be concerned with the internal representations of such structures in the computer, and their suitability for the different operations to be performed on them.

### 4.1.1 Definitions

An information structure consists of a set of nodes (sometimes called items or elements). The information at a node may be represented in one computer word or in several words. Each node can be subdivided into fields which again may require different amounts of storage for their representation, e.g. a bit, a byte, a word or several words.

The address of a node is the memory location of its first word. This is called the link. Alternatives to the word link are pointer (as used in PL/I) or reference (as used in Algol 68 and Algol W). The address of a node should not be confused with the value of the node which is the contents of the word or words starting at the address of node. The address may be the absolute address or more usually an address relative to some base address.

### 4.1.2 Basic Operations on Information Structures

The most important basic operations that we need to perform on data structures are:

(1) **Accessing.** We often want to obtain access to a node in a given position in the structure in order to examine its contents or to alter them.

(2) **Inserting.** Sometimes new nodes must be inserted into an information structure; for example, the birth of a child will cause new entries in a structure containing information about the whole family. Inserting a new node will normally involve finding the position in the structure where the new node has to be inserted and the difficulty of doing this will depend on the type of structure and how it is stored.

(3) **Deleting.** This operation is the inverse of inserting and naturally has many of the same characteristics.

(4) **Searching.** The problem here is to search the information structure for a particular node or nodes with a certain value in some field of the node. The field of the node being searched is called the key. This operation is, of course, similar to accessing but the identification of the required node by its contents instead of by its position raises quite different issues for most information structures. Searching is treated in some detail in chapter 7.

(5) **Sorting.** For some applications the nodes must be sorted into order according to a key kept in a particular field of the nodes. Chapter 8 describes several methods of sorting.

Other basic operations which are occasionally required are:

(6) **Copying.** The whole or a part of a structure may need to be copied into another structure.

(7) **Combining.** The establishment or recognition of some relationships between two or more structures can call for their combination into a single structure.

(8) **Separating.** This operation is the opposite of the previous one; a single structure is split into two or more structures.

In a particular application not all these basic operations are likely to be required, while some of the operations may be needed very frequently and others hardly at all. One of the biggest difficulties in choosing the best representation for an information structure is trying to balance several incompatible requirements. For example, a representation that was most efficient for accessing information in a certain structure would almost certainly be rather less efficient for deletion and insertion. We shall see that the best representations and algorithms for accessing information depend on whether the nodes are to be accessed in a systematic or random order.

### 4.1.3 Arrays

The idea of arrays of subscripted variables used in programming languages is taken straight from mathematical notation where we often refer to a set of related variables $x_1, x_2, \ldots x_n$ as the vector x. We use the subscript '1' or more generally 'i' to select a particular variable from the vector. The idea extends easily to variables with two subscripts. For example, we may consider $a_{ij}$ where the two independent subscripts are i and j which can take the values $i = 1, 2, \ldots m$ and $j = 1, 2, \ldots n$. Such a set of $m \times n$ variables can be represented as a two-dimensional array A (of which the matrix met in algebra is a special case).

$$A = \begin{pmatrix} a_{11} & a_{12} & \cdots & a_{1j} & \cdots & a_{1n} \\ a_{21} & a_{22} & \cdots & a_{2j} & \cdots & a_{2n} \\ \vdots & & & \vdots & & \vdots \\ a_{i1} & a_{i2} & & a_{ij} & & a_{in} \\ \vdots & \vdots & & \vdots & & \vdots \\ a_{m1} & a_{m2} & & a_{mj} & & a_{mn} \end{pmatrix} \qquad (4.1)$$

The first subscript is conventionally the row index and the second subscript the column index so that the variable $a_{ij}$ is in the ith row and jth column. The matrix represented above has m rows and n columns.

In theory any number of subscripts can be allowed to define a structure of this kind and we can even consider the simple variable or scalar as one with no subscripts. Most high level programming languages provide for these basic mathematical entities where the structures are simply called arrays. Each subscript of the array is called a dimension; thus a matrix like A would be an array with two dimensions. The most important arrays are the one-dimensional array (sometimes called a vector or table) and the two-dimensional array (or matrix). In most of this chapter the arrays of 'orthogonal information structures' will be considered to have their elements stored in consecutive storage locations of the same type. We will also study how the storage of an array (notably a two-dimensional array) needs to be affected by whether the array is dense or sparse. The effects of using external storage for arrays will also be considered.

## 4.2 STORAGE OF ARRAYS

Consider first arrays stored so that the elements occupy contiguous locations in storage. In Algol, Fortran and PL/I the storage is determined at the declaration stage. The following is a typical Algol declaration

**begin array** $x[k:l]$

which specifies that an array whose name is x is one-dimensional and that the single subscript of each element lies in the range k to $l$ inclusive where both k and $l$ are whole numbers (i. e. type **integer**). In some node or nodes we need to keep this information and also the address of the first element of x (i. e. x[k]). All this information and perhaps some other (e. g. whether x is real or integer) is kept in what is usually known as the <u>dope vector</u>. For arrays of higher dimensions the dope vector needs to contain the upper and lower bounds of each dimension. An array element can be accessed during execution of a program by abstracting the information in the dope vector and using it together with the subscripts of the element to calculate, by means of a mapping function, the particular address of that array element.

### 4.2.1 The Mapping Function

The mapping function $M(i)$ for a one-dimensional array is very simple. For the array $x$ declared previously, assume that the first element $x[k]$ is stored in the base address, BA. Then a general element $x[i]$ is stored in the address

$$M(i) = BA + i - k \qquad (i = k,\ k+1,\ \ldots\ l)\ .$$

Consider now the two-dimensional array declared as follows

**begin array** A[1:m, 1:n]

If this matrix A is ordered by rows then it would be stored in the order

$A[1, 1], A[1, 2], \ldots A[1, n], A[2, 1], \ldots, A[2, n], \ldots A[m, 1], \ldots A[m, n]$.

Again assuming the first element $A[1, 1]$ is stored in the base address, BA, the address of $A[i, j]$ is given by the mapping function

$$M(i, j) = BA + n(i - 1) + j - 1 \ldots . \qquad (4.2)$$

If the matrix A is ordered by columns then a different mapping function is needed (see example [4.2]). Thus, the mapping function specifies the allocation of storage to the array and the ordering of the array elements. During the execution of a program each access of an array element calls for an evaluation of the mapping function. Notice that the evaluation of (4.2) for accessing an element of a matrix requires only one multiplication and two additions if we store in the dope vector the value of $(BA - n - 1)$ instead of the Base Address itself.

For a three-dimensional array declared as follows

**begin array** C[1:$l$, 1:m, 1:n]

suppose we store this array so that the first subscript varies most slowly and the last subscript most quickly. The order in storage is, therefore

C[1, 1, 1], C[1, 1, 2],... C[1, 1, n], C[1, 2, 1],... C[1, m, n-1], C[1, m, n], C[2, 1, 1],... C[$l$, m, n].

The ordering is lexicographical and is sometimes called 'row major order'. The mapping function for the address of C[i, j, k] (assuming C[1, 1, 1] is in address BA) is

$$M(i, j, k) = BA + mn(i-1) + n(j-1) + k-1 \ldots \qquad (4.3)$$

Equation (4.3) illustrates the general rule for constructing a mapping function of n dimensions from one of n - 1 dimensions. We assume an n-dimensional array with general bounds on the subscripts:

**begin array** $D[l_1:u_1, l_2:u_2, \ldots, l_n:u_n]$ (4.4)

If D is stored in lexicographical order with $D[l_1, l_2, \ldots l_n]$ in address BA, the mapping function for $D[i_1, i_2, i_3, \ldots i_n]$, where $l_K \leq i_K \leq u_K$, is

$$\begin{aligned} M(i_1, i_2, \ldots i_n) = BA + ((\ldots (i_1 - l_1) \times (u_2 - l_2 + 1) + i_2 - l_2) \\ \times (u_3 - l_3 + 1) + i_3 - l_3) \times \ldots) \times (u_{n-1} - l_{n-1} + 1) + i_{n-1} - l_{n-1}) \\ \times (u_n - l_n + 1) + i_n - l_n \ldots \end{aligned} \qquad (4.5)$$

If we let $d_i = u_i - l_i + 1$ for $i = 1, 2, \ldots n$ so that $d_i$ is the 'thickness' of the ith dimension, and

$$F = BA + (((\ldots (-l_1) \times d_2 - l_2) \times d_3 + \ldots) \times d_{n-1} - l_{n-1}) \times d_n - l_n$$

then (4.5) can be rewritten

$$\begin{aligned} M(i_1, i_2, \ldots i_n) = ((\ldots (i_1 \times d_2 + i_2) \times d_3 + \ldots) \times d_{n-1} + i_{n-1}) \times \\ d_n + i_n + F . \end{aligned} \qquad (4.6)$$

Since F depends only on the subscript bounds $u_i$, $l_i$ and the base address it is a constant which can be evaluated once and for all when the array is declared. Thus, the evaluation of the mapping function given by (4.6) requires only (n - 1) multiplications and n additions.

In some languages (e. g. Fortran and Pascal) the dimensions of the arrays are fixed at declaration time. This simplifies the Base Address in the dope vector and the mapping function and thus array access is quicker. It has however the disadvantage that efficient array storage is usually not possible.

**Triangular, Symmetric, and Skew-symmetric Arrays.** We can often use a more complex mapping function to save memory space at the expense of added execution time. For example we sometimes need only the elements of a matrix A[i, j] which lie on or below the leading diagonal; symmetric matrices for which A[i, j] = A[j, i] and skew-symmetric matrices for which A[i, j] = -A[j, i] are particular examples. We can therefore save storage by storing just the lower triangular matrix A[i, j] with $j \leq i$, where the array has bounds A[1:n, 1:n]. The array can be stored lexicographically as follows:

A[1, 1], A[2, 1], A[2, 2], A[3, 1], A[3, 2], A[3, 3], . . . A[n, 1], A[n, 2], . . . A[n, n].

If A[1, 1] is stored in the base address BA then the mapping function for the address of A[i, j] is

$$M(i,\ j) = BA + \frac{i(i-1)}{2} + j - 1\ . \tag{4.7}$$

Although a few language compilers or interpreters include special facilities for certain shapes of arrays, such a mapping function is not usually provided and so the programmer must write the mapping expression explicitly himself. For the example above this could be done by declaring the **array** T [1:n × (n+1)/2]; and whenever the element A[i, j] is required in the program we use T[i × (i-1)/2 + j]. Thus we convert our matrix A into the vector T.

### 4.2.2 Access Tables

The mapping function method of storage can be inefficient if references to the elements of the array are frequent and if multiplication time is important, so that the work in calculating the address of the elements is significant. An alternative method is to use subsidiary data structures called access tables (or access vectors) which not only save doing any multiplication but are in many applications much more flexible than mapping functions. For example, they allow a relaxation of the requirement of contiguous storage locations for the entire array.

Consider the simple two-dimensional array A[1:3, 1:4] and suppose the first element A[1, 1] is stored in location 1000. The array elements and their storage locations are:

| | | | | |
|---|---|---|---|---|
| | A[1, 1] | A[1, 2] | A[1, 3] | A[1, 4] |
| Location | 1000 | 1001 | 1002 | 1003 |
| | A[2, 1] | A[2, 2] | A[2, 3] | A[2, 4] |
| Location | 1004 | 1005 | 1006 | 1007 |
| | A[3, 1] | A[3, 2] | A[3, 3] | A[3, 4] |
| Location | 1008 | 1009 | 1010 | 1011 |

We can see that the address of any element in the first row, A[1, j], is 999 + j, that of any element in the second row, A[2, j], is 1003 + j and similarly for any element in the third row, A[3, j], 1007 + j. Thus, if we have an access vector AV[1:3] whose three elements are respectively 999, 1003, 1007 we can access A[i, j], by the location AV[i] + j. This latter formula needs no multiplication. The essence of access tables is to obtain greater speed of access at the expense of more storage space. In general the extra storage space for an access table is small; for example for a two-dimensional array A[1:m, 1:n] we require $m \times n$ locations for the array plus m locations for the access vector, so that

$$\text{the total number of locations} = m(n + 1) . \qquad (4.8)$$

From (4.8) we can see that the space for an access table is less if the smaller dimension is first. Hence, for 'long, thin arrays' it is best to have the 'thin' dimension in the first subscript position.

For a square matrix, (4.8) gives the total number of locations to be $n(n + 1)$ which is less than $(n + 1)^2$. So in this case the use of an access table merely reduces the allowable maximum sized matrix by one.

We can generalize the access table method given above from a matrix to an n-dimensional array. Consider the array declared previously (4.4). The array, D, has dimensions $d_1 \times d_2 \times \ldots \times d_n$. The address of the element $D[i_1, i_2, \ldots i_n]$ is given by

$$AT1[i_1] + AT2[i_2] + \ldots + i_n$$

where AT1 occupies $d_1$ storage locations and gives the address (incorporating the base address) of the first element of the $(n - 1)$-dimensional 'slice' (i. e. $D[i_1, l_2, l_3, \ldots l_n]$).

We have to refer to the store for each of $AT1[i_1]$, $AT2[i_2]\ldots$, and so these arrays need other locations specifying their base addresses. If, for example, we store D first then AT1, AT2, ...

$D[l_1, l_2, \ldots l_n]$ is stored in base address (BA)
$D[u_1, u_2, \ldots u_n]$ in $BA + d_1 d_2 \ldots d_n - 1$
$AT1[1]$ in $BA + d_1 d_2 \ldots d_n$

⋮

$AT1[i_1]$ in $BA + d_1 d_2 \ldots d_n + i_1 - 1$

⋮

$AT2[1]$ in $BA + d_1 d_2 \ldots d_n + d_1$ .

⋮

Thus when the array is declared with known dimensions we can calculate the starting addresses of AT1, AT2, ... and we can reference $AT1[i_1]$ as $i_1$ + constant, where we have a store reference for the constant.

It is possible but not common to adopt strategies intermediate between the mapping function and the access table.

We will now look again at the problem of accessing a triangular array but solving it by using an access table instead of the previous mapping function. Consider the $n \times n$ array A and assume we only wish to access the lower triangle. As before the vector T contains the information in the lower triangle of A and we also need an access vector AT:

**array** T[1:n × (n + 1)/2]; **integer array** AT[1:n];

at the beginning we will execute the following loop which fills the access vector AT with the appropriate values (BA is the base address of T[1]):

AT[1]:=BA - 1; **for** i:=2 **step** 1 **until** n **do** AT[i]:=i + AT[i - 1] - 1;

This ensures that AT[i] = BA - 1 + i(i - 1)/2 for i = 1, 2, . . . , n. When we wish to use the value of A[i, j] we access the address AT[i] + j.

An alternative method of storing triangular arrays is to pack two such arrays into an $n \times (n + 1)$ rectangular array.

For example, consider the two lower triangular arrays A and B and n = 3 they can be packed into the array C as follows

$$C = \begin{pmatrix} A[1,1] & B[1,1] & B[2,1] & B[3,1] \\ A[2,1] & A[2,2] & B[2,2] & B[3,2] \\ A[3,1] & A[3,2] & A[3,3] & B[3,3] \end{pmatrix}$$

thus $A[i, j] = C[i, j] \qquad i \geq j,$

$B[i, j] = C[j, i+1] \qquad i \geq j.$

An interesting, although not commonly encountered problem is the extension of this idea to three dimensions - packing tetrahedral arrays into a box-shaped one.

## 4.3 APPLICATIONS OF ACCESS TABLES

1. **Row interchange.** The interchange of two rows of a matrix is an important problem in programming matrix calculations. For example, many of the better algorithms for solving large sets of simultaneous linear equations require row interchange to prevent small pivots giving large round-off errors. The use of access tables gives us a very efficient way of doing the row interchanges. If the matrix elements A[i, j] are addressed using AT[i] + j then in order to interchange the whole rows s and t all we need to do is interchange the single entries AT[s] and AT[t] in the access table. The rows s and t are in effect interchanged without physically moving any of the elements of the matrix A.

2. **Jagged arrays.** Jagged arrays are a generalization of the arrays previously considered in that the 'slices' of an array (i. e. the

rows in the case of a matrix) may contain different numbers of elements. If access tables are used this new concept is easily implemented since only the contents of the access tables need to be changed. An example of the use of a jagged array is the manipulation of English text in a computer. Each element of the array could be a letter and the full array TEXT could be

TEXT [SENTENCE, WORD, LETTER] .

If jagged arrays are not used we must use large upper limits for the number of letters in a word and the number of words in a sentence. We would have a lot of wasted, unused storage space and even so an unanticipated long word or long sentence could cause overflow. The access table method of storing the jagged array means that the whole text can be stored without any wasted space.

3. **Sorting with detached keys** (see also section 8.1). Suppose we have items which occupy several words of storage and we wish to sort them into order according to the key field.

Let the ith item consist of Key i Data i and be stored in the words starting at location $D_i$. In store we have, therefore,

| Location | | |
|---|---|---|
| $D_1$ | Key 1 | Data 1 |
| $D_2$ | Key 2 | Data 2 |
| ⋮ | ⋮ | ⋮ |

If we physically rearrange all the data it would require a large number of data transfers especially if the data part is lengthy. A much more efficient method is to form an access table:

| | | |
|---|---|---|
| Key 1 | : | $D_1$ |
| Key 2 | : | $D_2$ |
| ⋮ | | ⋮ |

and then interchange the items in this access table until they are in the correct order.

4. **Use of pockets of storage.** In some language processors and other programs, storage is allocated and released during execution so that non-consecutive pockets of store are available at various times. These pockets of storage can be allocated to array structures without much difficulty if access tables are used.

5. **External storage.** Many problems require large arrays and only part of such an array may be stored internally at any one time. The rest of the array would be stored on some external storage such as a magnetic disc or a magnetic tape, while the slices of the array currently in use are stored in the high speed store. We can use an access table for such a large array and let the elements of the access table which correspond to the slices in high speed store point to those rows, while the other elements of the access table can indicate that a slice is in external storage and cause an interrupt to occur. The interrupt would allow an administrative program to fetch the missing slice into high speed store, removing another slice or slices if necessary and updating the access vector to reflect this new state of affairs. Then finally, the program would be able to access the new slice of the array brought in.

## 4.4 SPARSE ARRAYS

An important particular case of the general orthogonal array which requires special methods of storage and manipulation is the sparse array. A sparse array is one in which most of the elements are zero and, while it is not a precisely defined term, one would say an array was sparse if more than 90 per cent of the elements were zero. An array is treated as 'sparse' if it is worthwhile adopting special methods for its storage and manipulation. Many practical problems involve large matrices most of whose elements are zero. For example, matrices arising from framework problems or from differential equations. In such cases it is usually not possible to store all the elements and very inefficient to operate on all the zero and non-zero elements. We are, therefore, looking for algorithms which only store and operate on the non-zero elements. One way to do this, which is efficient when the elements of the array are accessed in random order, is to use hash addressing techniques by methods which are explained in section 7.3. However, most array manipulation requires the elements to be accessed in certain orders dependent on their structures and the operations required. The hash addressing methods are inappropriate in these cases. We will now examine various methods of storing a non-zero value of a sparse array together with some method of determining its position.

### 4.4.1 Binary Pattern

A binary pattern is formed which indicates the position of the non-zero elements by making a '1' for each while '0's denote the positions of zero elements. The values of the non-zero elements are stored sequentially after the binary pattern. For example, the array

$$\begin{pmatrix} 0 & 7 & 0 & 0 & 0 \\ 1 & -2 & 0 & 0 & -3 \\ 0 & 0 & 4 & 0 & 0 \\ 12 & 0 & 0 & 0 & 0 \end{pmatrix} \tag{4.9}$$

will be stored as the binary pattern

$$\begin{pmatrix} 01000 \\ 11001 \\ 00100 \\ 10000 \end{pmatrix}$$

followed by the values 7; 1; -2; -3; 4; 12;

In order to use this method effectively the binary patterns must be packed efficiently into the computer store. For example, if the array is a two-dimensional $m \times n$ matrix with $N$ non-zero elements then the amount of storage needed is approximately

$$N + \frac{m.n}{W}$$

where $W$ is the number of bits per word. Many high level language compilers store their bit patterns as logical variables which each occupy a word or a byte. If the storage is of this type then the binary pattern method is of little use.

Access to a given position in the array is by using the bit pattern storage markers, masks and the logical 'and'. If the element is non-zero its value is extracted from the sequence after counting the previous 1's. It is often advantageous to store row markers and also to allow some bits to store a count of the number of 1's so far. Even so there is some penalty in accessing for the gains obtained in storage.

Insertion and deletion are awkward with this representation of a sparse array. The alteration of the bit pattern is not too difficult but the inserting of the new value in the correct position in the sequential list of values is awkward.

### 4.4.2 Co-ordinates

One of the most commonly used methods of manipulating a sparse array is to store in lexicographical order only the non-zero values together with their co-ordinates. For example, the array (4.9) can be stored as

1; 2; 7; 2; 1; 1; 2; 2; -2; 2; 5; -3; 3; 3; 4; 4; 1; 12;

and, for a general n-dimensional array the non-zero element

$A[i_1, i_2, \dots i_n]$

is stored $i_1; i_2; \dots; i_n;$ value of $A[i_1, i_2, \dots i_n]$.

Consider the special case of a matrix stored in this way and suppose we wish to access the element with subscripts x, y. Let there be N non-zero elements stored in the array **store** [1: 3 × N] and that each of the N values is preceded by its 2 subscripts. A possible piece of program is

```
for s:= 1 step 3 until 3 × N do
begin i:= store [s]; j:= store [s + 1];
        if i = x and j = y then goto FOUND;
    if i > x or i = x and j > y then goto NOT FOUND;
end;
```

When we jump out to the label FOUND the element with the subscripts x, y has the value **store** [s + 2]. If we jump out to the label NOT FOUND the element with the subscripts x, y is a zero value.

Inserting and deleting are difficult although in the latter case we can just make the value zero and leave it stored. This does not save storage or processing time but does avoid reorganizing the stored elements on every occasion of deletion. If insertion is an important operation in the manipulation of the arrays then this method is unsuitable. Either the new element is inserted in its correct position and we have to rearrange many of the stored elements or the new element is inserted at the end and the array elements are unordered and only less efficient algorithms are available.

There are many variations on this method of storing sparse arrays, see for example exercises [4. 12], [4. 13], and [4. 14].

### 4. 4. 3 Co-ordinates and Linked Allocation

The previous method of storing sparse arrays can be modified to make insertion of new elements easy by adding extra items of information for each element. The extra information is the locations of the neighbouring items. In the simplest case this consists of a list with each element represented by its subscripts, its value, and the address of the next element in the list.

Consider the sparse array (4. 9) and suppose that originally the elements, and their co-ordinates, are stored in locations 100-123. This array would be stored, therefore, as in fig. 4. 1, where -999 is used to

| Location | Value | Location | Value |
|---|---|---|---|
| 100 | 1 | 112 | 2 |
| 101 | 2 | 113 | 5 |
| 102 | 7 | 114 | -3 |
| 103 | 104 | 115 | 116 |
| 104 | 2 | 116 | 3 |
| 105 | 1 | 117 | 3 |
| 106 | 1 | 118 | 4 |
| 107 | 108 | 119 | 120 |
| 108 | 2 | 120 | 4 |
| 109 | 2 | 121 | 1 |
| 110 | -2 | 122 | 12 |
| 111 | 112 | 123 | -999 |

Fig. 4. 1

indicate a null pointer and thus the end of the list. If we wish to insert an element with subscripts 3, 1 and value 17 we store these values in the next three available locations say 124, 125 and 126, in location 127 we store 116 and location 115 is altered from 116 to 124.

Items are not, therefore, sequential in storage and we access items starting at the list head and progressing via the links. The ideas used in linked storage allocation will be discussed in more detail in the next chapter; it is already clear that greater ease of inserting elements is being obtained at the expense of more store and more work to access items.

The method described above with one link for each element is fairly rudimentary and in order to manipulate sparse matrices in the operations of transposition, addition, multiplication, inversion, row or column interchange, etc. more complex linkages are sometimes used. One such method is described in some detail by Knuth (vol. 1) where in addition to the subscripts and the element value two links are stored, one of which is to the previous non-zero element in the row and the other to the non-zero element above it in the column.

## 4.5 BIBLIOGRAPHY

Most books on programming discuss arrays and array storage. One of the earliest and still a most useful document is the Carnegie Institute of Technology Monograph by Braden and Perlis [1]. Another important source for further material on the topics of this chapter is Knuth, vol. 1 [2]. In particular the reader interested in the storage of sparse arrays using row and column linkage will find it explained in Knuth together with an algorithm for the pivot step in matrix inversion.

Pooch and Nieder [3] have recently produced a good general survey of methods of storing and indexing sparse matrices. There are also four relatively recent books on Data Structures by Berztiss [4], Elson [5], Shave [6], and Stone and Siewiorek [7], which include sections on arrays but are also very relevant to the subsequent chapters.

[1] Robert T. Braden and Alan J. Perlis: An Introductory Course in Computer Programming, pp. 35-55. D-S-C Monograph no. 7, Carnegie Institute of Technology, 15th June, 1965.

[2] Donald E. Knuth: The Art of Computer Programming, vol. 1: Fundamental Algorithms, pp. 295-304. Addison-Wesley, 1968.

[3] U. W. Pooch and A. Nieder: A survey of indexing techniques for sparse matrices, ACM Computing Surveys, vol. 5, no. 2, pp. 109-33, June 1973.

[4] A. T. Berztiss: Data structures - Theory and Practice, 2nd Edition, Academic Press, 1975.

[5] Mark Elson: Data Structures, Science Research Associates, 1975.

[6] M. J. R. Shave: Data Structures, McGraw-Hill, London, 1975.

[7] H. S. Stone and D. P. Siewiorek: Introduction to Computer Organization and Data Structures: PDP-11 Edition, McGraw-Hill, New York, 1975.

## EXAMPLES 4

[4.1] Given the following arrays stored in lexicographical (or row major) order derive the appropriate mapping functions

(a) **array** A[0:m; 2:n];

(b) **array** B[-1:m, 0:n, 5:p, -q:r];

[4.2] Given the following arrays stored by columns (or reverse lexicographical order) derive the appropriate mapping functions

(a) **array** C[1:m, 1:n];

(b) **array** D[-1:m, 0:n, -p:q];

(c) **array** $E[l_1:u_1, l_2:u_2, \ldots, l_n:u_n]$;

(d) Repeat part (a) above but allow each element of the array to occupy three words of storage.

[4.3] A tridiagonal array A of size $n \times n$ is a matrix in which only the leading diagonal and the two adjacent ones contain non-zero elements. The $3n - 2$ non-zero elements of A are to be stored compactly in the linear array T and accessed via the **integer array** AT[1:n]; such that A[i, j] is accessed at T[AT[i] + j]. Determine the values that must be put in the access vector AT.

[4.4] Repeat example [4.3] for a banded matrix A whose bandwidth is $(2k + 1)$, i.e. $A[i, j] = 0$ if $|i - j| > k$.

[4.5] Describe how pointers are used in information processing, with particular reference to the storage of arrays. What are the advantages of this method of storing an array?

(Southampton (Eng.) 1969)

[4.6] An array a is lower-triangular, having zero elements above the main diagonal, i.e. $a_{ij} = 0$ for $j > i$. A FORTRAN program has been written to manipulate this array assuming that all the elements have been stored (including the zeros). In order to conserve storage space it is proposed to store only the non-zero elements of the lower triangle in the linear array AVEC, and to define a function A such that A(I, J) has the value of the array element $a_{ij}$. Write a FORTRAN function segment to meet this specification.

A certain $n \times n$ matrix is known to be sparse, with only 10 per cent of its elements having non-zero values. However, the non-zero elements are not distributed in any systematic way. Devise a way of storing this array so as to minimize the storage space occupied, and outline the programs that would be needed to access the array.

(Southampton, Diploma, 1969)

[4.7] Compare and contrast the storage of square matrices using a mapping function or using access tables.

Given an upper triangular array x[i, j], where $1 \le i \le j \le n$, we wish to store the elements in consecutive locations in the order x[1, 1], x[1, 2], ... x[1, n], x[2, 2], x[2, 3], ... x[n, n]. If x[1, 1] is stored in a base address, $A_0$, give the mapping function for the array x. How would you store this array using access tables? In what situations are the use of access tables particularly valuable?

(Newcastle 1970)

[4.8] An array is said to be sparse if only a relatively small number of its elements have a value other than zero. The wastage of store resulting in this case from the usual representation of an array may be avoided by

(a) storing only the non-zero elements, and their position, in some data structure, and

(b) providing some mechanism which uses this structure to

find or change the value of an element.

What more economical data structure would you adopt in place of a large sparse one-dimensional array of real numbers, when the elements are to be referred to many times but are not to be changed in value?

Write an efficient ALGOL procedure declaration to find the value of an element specified in terms of its position in the original array. Explain your method and any parameters needed.

(Leeds 1971)

[4.9] Describe a method of storing an $(n \times n)$ sparse matrix which is economical in storage space, and give an algorithm for accessing an element of the matrix given its subscripts. How would you insert a new element into the matrix?

Two $(n \times n)$ matrices A and B have respectively x and y non-zero elements. How many non-zero elements would you expect to appear in $A + B$ and $A \times B$? You may assume that both x and y are small in comparison with $n^2$ and that the non-zero elements appear in a random manner in A and B.

(Newcastle 1971)

[4.10] Describe a method for representing, in core storage, a two-dimensional array, the rows of which are of unequal length. Provision should be made for preventing access to non-existent elements. Indicate how, at the level of machine code, an element is accessed, after validation. What economy can be effected if a number of similar arrays (i. e. of the same shape) are to be represented? Outline a method for representing two-dimensional sparse arrays which avoids storing the zero elements.

(Essex 1971)

[4.11] Compare and contrast the storage of square matrices using mapping functions or using access tables. Given the following Algol W declaration

```
REAL ARRAY X(0 :: N, 0::N);
```

what would the appropriate mapping functions and access tables be if this matrix is stored (a) by rows, (b) by columns? Assume that X(0, 0) is stored in the base address BA.

An upper Hessenberg matrix a(i, j) is one which consists of the upper triangle and the diagonal below the leading diagonal; i. e. a(i, j) is zero if $i - j \geq 2$. What would be a suitable mapping function for this matrix if we wish to store the elements for which $i - j < 2$ in consecutive locations in the order

a(1, 1), a(1, 2), a(1, 3), ..., a(1, n), a(2, 1), a(2, 2), ...
a(2, n), a(3, 2), a(3, 3), ..., a(3, n), ..., a(k - 1, k),
a(k, k), ..., a(k, n), ..., a(n - 1, n), a(n, n)?

Assume that a(1, 1) is stored in the base address BA. How would you store this matrix using an access table?

(Newcastle 1973)

[4.12] An $(n \times n)$ sparse matrix with m non-zero elements is stored as three one-dimensional arrays ROW, COLUMN and VALUE.

ROW has bounds 1 to n.
COLUMN and VALUE have bounds 1 to m.

COLUMN (J) and VALUE (J) are respectively the column position and value of the J-th non-zero element. ROW (I) on the other hand is a pointer to the first non-zero element in the I-th Row, e. g. if ROW (4) = 6 then the first non-zero element in the 4-th row is in COLUMN (6) and has VALUE (6). As an example the $4 \times 4$ matrix

| | | | |
|---|---|---|---|
| 5 | 0 | 3 | 0 |
| 2 | -1 | 5 | 0 |
| 0 | 0 | 0 | -2 |
| 1 | 4 | 0 | 0 |

would have the following values in ROW, COLUMN, and VALUE.

| ROW (1) | ROW (2) | ROW (3) | ROW (4) |
|---|---|---|---|
| 1 | 3 | 6 | 7 |

| | 1 | 2 | 3 | 4 | 5 | 6 | 7 | 8 |
|---|---|---|---|---|---|---|---|---|
| COLUMN | 1 | 3 | 1 | 2 | 3 | 4 | 1 | 2 |
| VALUE | 5 | 3 | 2 | -1 | 5 | -2 | 1 | 4 |

Write an algorithm which would return the value of the element in the r-th row and s-th column of the original sparse matrix. How would you insert new elements?

An $(n \times n)$ sparse matrix A has p non-zero elements. How many non-zero elements would you expect to appear in $A^3$? You may assume that p is small in comparison with $n^2$ and that the non-zero elements appear in a random manner in A.

(Newcastle 1975)

[4.13] Explain briefly how sparse matrices differ from full ones and how advantage can be taken of this difference. Explain, in very general terms, how you would set about solving a system of linear equations whose coefficient matrix was sparse.

Consider the matrix with the structure (column pointer/row index format):

IP: 1 4 6 8 9
IRN: 1 2 3 1 4 2 3 1

then write down the two relevant arrays for storing the same structure in a row pointer/column index format. Gaussian elimination is performed on this matrix using modified Markowitz as a criterion for pivot selection.

Give a possible pivot order, fill-in, and multiplication count. What would the fill-in be if pivoting down the diagonal in the natural order were used?

Consider the matrices with the structures (column pointer/row index format):

(i) IP: 1 3 6 8 9 12 15 17
IRN: 1 4 2 5 6 1 7 5 1 3 7 2 3 6 4 7

(ii) IP: 1 2 4 6 7 9 12
IRN: 3 3 5 2 4 3 1 3 2 4 6

Which matrix is structurally singular and what is the length of its transversal? Give a column permutation which puts non-zeros on the diagonal of the other one.

(Newcastle 1974)

[4.14] Sparse matrix programming projects

A sparse matrix is one with a large proportion of zero elements; if advantage is taken of this property, the amount of storage required for the representation of such a matrix can often be greatly reduced. The objectives for this group of projects are to measure and compare the benefits of a number of techniques that have been suggested for sparse matrix representation. The measurements are to be made in the context of the Gauss elimination algorithm (with row interchanges) for the solution of a set of simultaneous linear equations. For each project, the student should construct a program which will read test data, set up the appropriate internal representations, and then execute and measure the performance of the Gauss elimination algorithm using this representation.

Measurements should take into account the storage required for the particular representation, the processing time required for the elimination algorithm, and some measure of the complexity of the program required to implement the algorithm with this representation. These measurements should be made over a range of test-cases, the collection or generation of which could be carried out cooperatively by the group.

During the elimination process, the original matrix is subjected to a sequence of transformations; non-zeros below the diagonal are gradually eliminated, while other zero elements may become non-zero. The net result may well be an overall increase in the storage required, and this 'fill-in' should be investigated. It may also be useful to analyse the back-substitution phase of the algorithm separately, as this does not involve any updating of the matrix. Another point to consider is the strategy for selection of pivots. Apart from the recommended practice of selecting pivots of maximum modulus, it is also possible to select pivots on the basis of the expected amount of 'fill-in' that will result from their choice.

Sparse matrix representations

Apart from the values of the non-zero elements, a sparse matrix representation must also include some indexing information to provide access to specific elements $a_{ij}$, given i and j. In the extreme case (referred to as method 0 below), we hold all elements, zero or not, and use the conventional address mapping provided by the AlgolW array subscript mechanism. Most schemes that have been devised use some combination of the following items of indexing information for an element $a_{ij}$:

(i) 1-bit code, 0 for zero ($a_{ij}$ not stored), 1 for non zero;

(ii) row index (i);

(iii) column index (j);

(iv) pointer to next non-zero element in row;

(v) pointer to next non-zero element in column;

(vi) (vii) backward pointers corresponding to (iv), (v).

The choice of representations given below is not intended to be exhaustive; many variations and combinations will suggest themselves; and possibly some new ideas. In general, methods using linked lists should perform well during the elimination phase, when new non-zeros have to be introduced. It is assumed throughout that the matrices to be represented are general, non-singular and non-symmetric.

The best general reference is Pooch and Nieder's survey paper referenced in Section 4.5. Another useful survey is I. S. Duff, A Survey of sparse matrix research, Proc. IEEE (USA), vol. 65, no. 4, April 1977, pp. 500-35, which contains about 600 references.

[0. Full matrix storage. This is impractical for very large matrices, but provides a standard for comparisons.]

1. Bit-map methods. An $n \times n$ array of bits indicates which elements are non-zero; corresponding non-zero values are stored separately. One variation is to store the bit map and elements by columns rather than rows. (See Section 4.4.1, and also F. G. Gustavson et al., JACM 17 (1970), pp. 87-109, J. de Buchet, in Large sparse sets of linear equations, edited by J. K. Reid, Academic Press, 1971, pp. 211-7.)

2. Coordinate method. Row and column indices (i and j) and element values ($a_{ij}$) are stored in linear arrays; ordering the entries by increasing

values of i and then j may be useful. (See Section 4.4.2.)

3. Column (or row) arrays. For each column (or row), non-zero elements and their row (or column) indices are stored sequentially. Separate 'pointers' may indicate the start of a column (row). (See A. R. Curtis and J. K. Reid, J. IMA 8 (1971), pp. 344-53 and F. G. Gustavson, Some basic techniques for solving sparse systems of linear equations, in Sparse matrices and their applications, edited by D. J. Rose and R. A. Willoughby, Plenum Press, 1972.)

4. Column (or row) lists. Each column (row) is represented by a linked list of non-zero entries together with row (column) indices. Backward links may also be included. (See E. C. Ogbuobiri, IEEE Transactions on PAS, 89 (1970), pp. 150-5.)

5. Doubly-linked lists. The set of non-zero matrix elements are stored in a doubly-linked structure, with a chain of links along both rows and columns. (See Knuth, vol. 1, pp. 298-302.)

6. 'Hash-coding' methods. The idea here is to use the techniques of 'hash-coding' to carry out an efficient search for $a_{ij}$, given values of i and j. (See E. v. d. S. de Villiers and L. B. Wilson, Hashing the subscripts of a sparse matrix, BIT, vol. 14 (1974), pp. 347-58.)

7. Variable bandwidth method. For each column (or row), store the row indices of the first and last non-zeros in that column together with all element values (zero or not) within this 'band'. As the elimination proceeds, the bands may increase in width. (See A. Jennings, Comp. J. 9 (1966), pp. 281-5 which describes a scheme for symmetric matrices.)

# 5 · Information Structures 2: Linear Lists

## 5.1 INTRODUCTION

Linear Lists are one of the simplest type of information structure and include one-dimensional arrays as a special case. It is just a list of items which are ordered by a single criterion so that each item, except possibly the first or last, will have a predecessor and a successor. More formally a linear list is defined as a set of n nodes $(n \geq 0)$ x[1], x[2], ... x[n] whose structural properties essentially involve only linear (one-dimensional) relationships between the nodes. i.e. x[1] is the first node, x[n] is the last node, and the node x[k] is preceded by x[k - 1] and followed by x[k + 1].

The one-dimensional array is obviously a linear list. The matrix or two-dimensional array (4.1) can be considered as a generalization of the linear list. In that array each element a[i, j] of the matrix can be considered as belonging to two linear lists, the 'row i list' a[i, 1], a[i, 2], ... a[i, n] and the 'column j list' a[1, j], a[2, j], ... a[m, j]. The two-dimensional structure of the matrix is thus exhibited by the two orthogonal lists. Similar extensions can be made for multi-dimensional arrays.

The basic operations we wish to perform on linear lists are those described in section 4.1.2. In this chapter we will first discuss some important special cases of linear lists and then examine the suitability for the various operations of the sequential and linked methods of storage allocation.

## 5.2 STACKS, QUEUES AND DEQUES

The first and last nodes of a linear list often have a special significance for the most important operations of inserting, deleting and accessing; for convenience special names have been given to linear lists in which all insertions, deletions and access are at the first or last nodes. The names we shall use and alternatives that may be encountered are as follows:

Stack: All insertions, deletions and access are made at one end of the list. (Alternative names: nesting store, push-down list, last-in-first-out list, LIFO.)
Queue: All insertions are at one end and all deletions and access at the other end. (Alternative names: first-in-first-our list, FIFO.)
Deque: (double ended queue) Insertions, deletions and access can be made at both ends of the list, but at no other nodes.

Some illustrations of these types of linear lists are given in example [5.1].

### 5.2.1 Stacks

Of the special linear lists named in the last section the stack is one which appears in several different areas of computing science and application. In some machines (e.g. KDF 9 and B 5000) a special stack is provided in the hardware. The following model (fig. 5.1) of a stack was suggested by E. W. Dijkstra and is usually known as 'Dijkstra's shunting yard'.

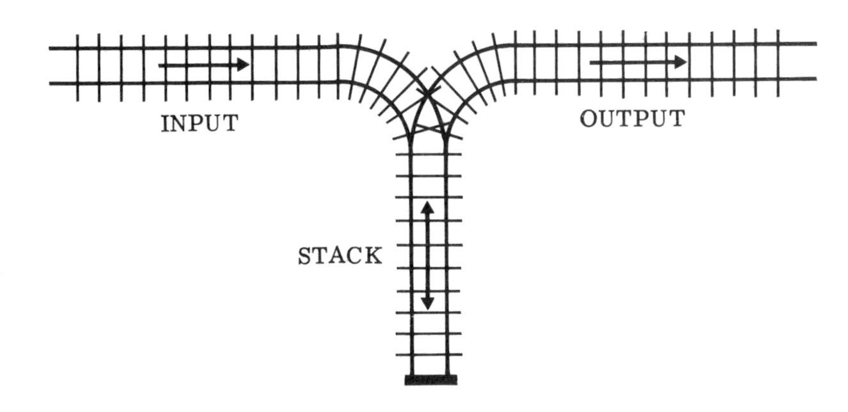


Fig. 5.1. Dijkstra's shunting yard

To illustrate the way in which a stack works consider the six data items A, B, C, D, E, F, on the input stream in that order and suppose that we wish them to appear on the output stream in the order C, B, D, E, F, A. The only primitive operations available are S (stack) which puts the next item in the input stream onto the top of the stack and U (unstack) which removes the top item of the stack and puts it on the output stream. Then the sequence SSSUUSUSUSUU will form the output stream CB D E FA where gaps denote the occurrence of S operations. It is convenient to think of a stack as a 'push-down'-'pop-up' list, so arranged that when we insert an item at the top of stack we push the others in the stack down. Similarly when we unstack the top item the others pop up.

Examples in other applications are the magazines of certain semi-automatic rifles in which cartridges and bullets are loaded by forcing down a spring which later feeds the last bullet entered into the breach, and the small coin containers often carried by motorists for a reserve of fodder for parking meters. However these models would be very inconvenient if we attempted to follow them for implementing a stack on a computer since they imply the motion of the stack in the computer store. Moving a large number of nodes in storage every stack operation is very inefficient. Instead a stack would be implemented in a computer by using a stack pointer T which points to the top node of the stack and moving the pointer instead of all the nodes of the stack. If the stack is empty, the stack pointer T = 0. The stack pointer advances when we place an item on the stack and retreats when we take the top item off the stack. This leads us to two important ideas in operating with linear lists overflow and underflow. Since our stack will not be infinite we must check to make sure it is not full before we add another item which would otherwise cause overflow. Also before we take an item out of the stack we must check to see that it is not empty (underflow).

Stacks occur very frequently in computing. In fact any type of problem, which has a subproblem, which requires the solution of a further subproblem and so on, has a stack like structure, since we can stack the subproblems as they occur and remove them from the stack when they are solved. The return address (or link) of a subroutine is an example. When the subroutine is activated we stack the return address; this subroutine may in turn invoke other subroutines and each time we stack the return address. When we leave any subroutine the top of stack will contain the correct return address for that subroutine. Languages with a nested structure can be conveniently processed using stacks. A typical example is the block structure of Algol. When the variables are declared in the blockhead the storage can be allocated as a stack and then the storage can be released in the correct order when leaving a particular block. Recursion and stacks have a natural connection; e.g. the maintaining of locally declared identifiers or parameters called by value in an Algol recursive procedure is most easily done using stacks. Stacks also occur in parsing arithmetic expressions when re-writing them from their printed form into Reverse Polish notation.

### 5.2.2 Queues

A stack requires only one pointer since all the operations occur at one end of it. However a queue requires one pointer for each end. Let these pointers be denoted F and R (standing for the front and rear of the queue respectively), and let the empty queue be represented by $F = R$ where initially both F and R are zero. The simple approach to queue storage is to have a set of storage locations and add to the rear and delete from the front as necessary. This has a serious drawback if the queue is not regularly emptied. For example, consider the case where initially there are 10 insertions into an empty queue and then eight deletions. Subsequently there are several series of ten insertions followed by ten deletions. At no time are more than 12 storage locations occupied by items. Suppose we have available the storage $x_1, x_2, \ldots x_n$. The straightforward approach requires stores $x_1, x_2, \ldots x_{10}$ for the first 10 insertions and then just stores $x_9, x_{10}$ after the 8 deletions. The next ten insertions cause stores $x_9, x_{10}, x_{11}, \ldots x_{20}$ to be occupied. The next ten deletions leave $x_{19}, x_{20}$, full and so on.

This method is therefore very wasteful of storage space unless the queue is regularly emptied when we can start afresh filling $x_1$. This difficulty can be avoided by considering the queue to be a circular list. The circular list can be allocated the nodes x[1], x[2], ... x[max], with x[1] considered to be after x[max] (see fig. 5.3).

Notice that as with stacks we have to guard against both overflow and underflow.

### 5.2.3 Deques

The most general linear list in which all accessing, deletions and insertions take place only at the ends is the deque. It can be represented diagramatically (fig. 5.2) in a similar way to the Dijkstra shunting yard for the stack where we have of course to imagine that the capacity of the deque (AB) is variable.

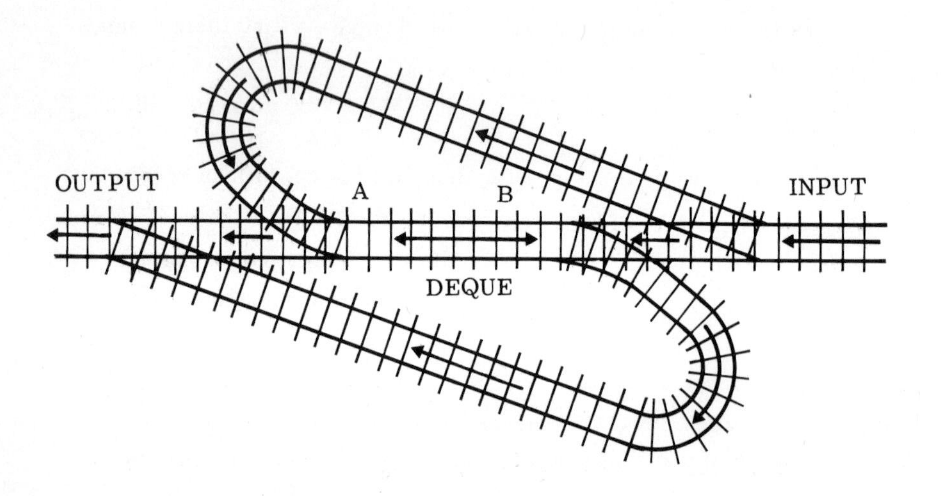


Fig. 5.2. A shunting yard representation of a deque

As with the queue we require two pointers F and R which point to the front and rear of the deque. The empty deque will be represented by F = R.

## 5.3 SEQUENTIAL ALLOCATION OF STORAGE

The simplest and most obvious way to store a linear list is to put the nodes in sequential storage locations in the computer. Thus if the nodes each occupy C words of storage per node then the location of x[k] is C words on from the location of x[k - 1]. For the linear list x[1], x[2], ... x[n] it would be normal to work relative to a base address which could be the address of the artificial node x[0]. The location therefore of x[k] is Ck words on from this base address. Throughout the rest of this section we will consider the case C = 1 which illustrates all the general features: simple alterations suffice when each node occupies more than one word of storage. In a later section we consider the various operations on linear lists when the nodes are not stored sequentially.

### 5.3.1 Stack Operations

The only two operations that can be performed are to load a new element onto the top of the stack and remove the top element from the stack. These operations will be written in an Algol-like notation to give more precision to the ideas used in these algorithms. Assume the stack contains the elements

x[1], x[2], ... x[T].

(i) To load the element y into the stack:

```
if T = Max then goto OVR;
T:= T + 1; x[T]:=y;
```

(ii) To remove the top element of the stack and put it into node y:

```
if T = 0 then goto STACK EMPTY
          y:= x[T]; T:= T - 1;
```

As explained in section 5.2.1 the empty stack is represented by a zero stack pointer (i.e. T = 0). 'Max' in the above piece of program is the maximum number of storage locations allocated to the stack. The labels or procedures OVR and STACK EMPTY can be either simple error messages or recovery routines. If the stack overflows this is usually an error caused by uncontrolled loading but it may arise from a genuine use requiring more than the normal space and then we may wish to allocate more storage locations to the stack if they are available. Attempting to remove items from an empty stack is often an error but it could also be used as a test for an empty stack.

### 5.3.2 Queue and Deque Operations

As explained in section 5.2.2 the queue is best represented as a circular list. The nodes x[1], x[2], ... x[max] will be allocated for this list, with the node after x[max] considered to be x[1]. Let the pointers to the front and the back of the queue be F and R respectively and the empty queue be represented initially by F = R = max. The only two operations we can perform on a queue are adding to the rear and removing from the front.

(i) To load the element y at the rear of the queue:

```
R:= if R = max then 1 else R + 1;
if R = F then goto QUEUE FULL; x[R]:= y;
```

(ii) To remove the element at the head of queue and put it into node y:

```
if R = F then goto QUEUE EMPTY;
            else begin F:= if F = max then 1 else F + 1;
                             y:= x[F];
            end;
```

In the algorithm (i) above the exit to the label QUEUE FULL takes place when there are max - 1 items in the queue. The reason for this is that it is difficult to distinguish between a full queue and an empty queue when all the storage space is fully utilized. Thus it is worth giving up one storage location to make the distinction between full and empty queues relatively simple in the algorithms. After a certain number of loadings and unloadings a typical queue would look like that shown in fig. 5.3.

Note that the rear pointer R points to the last element that was put into the queue, and the front pointer F points to the location <u>one ahead</u> of the next element to be removed from the queue.

A deque can be considered as an extension of a queue and thus it seems logical to represent it in the same way with pointers F and R to the front and rear of the deque respectively and the empty deque represented by F = R. The two primitive operations (i) and (ii) given previously for a queue will also hold for a deque and we need to add to further primitive operations to load at the front and delete from the rear.

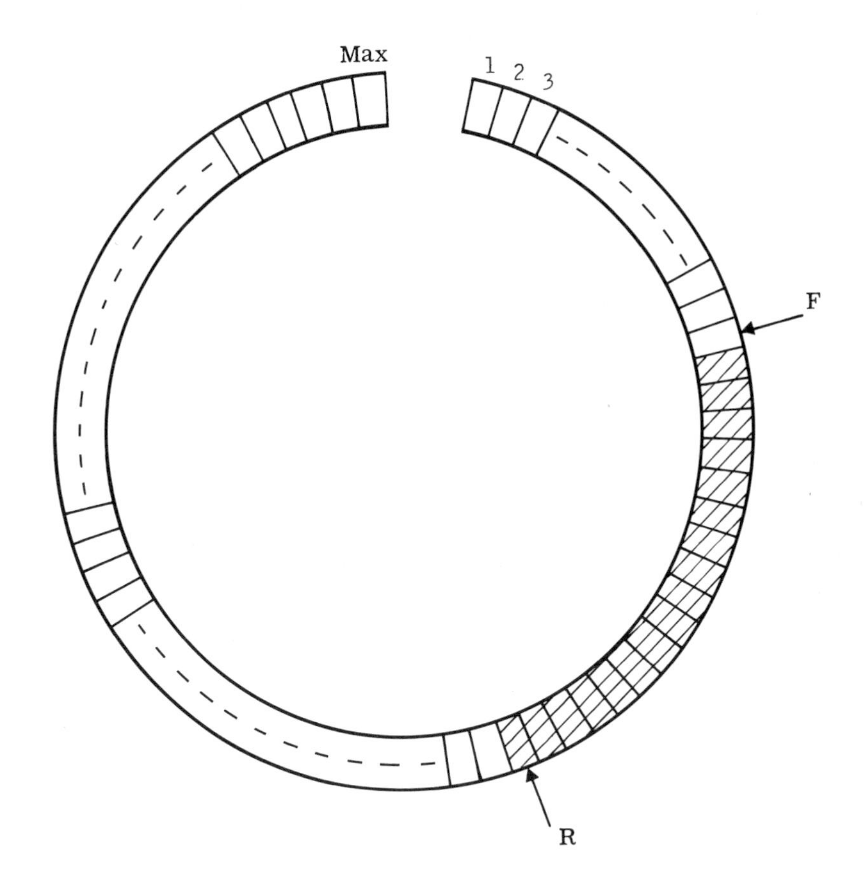


The actual locations occupied by the queue are shaded.

Fig. 5. 3. A queue or circular list

(iii) To load the element y at the front of the deque (this has been left as example [5. 5] for the reader).

(iv) To remove the element at the rear of the deque and put it in node y:

```
        if R = F then goto DEQUE EMPTY;
                  y:= x[R];
    R:= if R = 1 then max else R - 1;
```

### 5.3.3 Overflow and Lists

If we are manipulating several lists and have allocated each one a fixed storage space then we can find that one list has overflowed its allocation but there is plenty of unused storage space in the other lists. As the lists are unlikely to reach their maximum size at the same time this situation will occur quite frequently. We would like to have an algorithm that manipulated the storage requirements of the lists in such a way that overflow only occurred when all the available storage space was used. This is in general an exceedingly difficult problem and the only elementary results are for stacks. The simplest case is that of two stacks and this can be solved quite neatly by letting the two stacks grow towards each other, as shown in fig. 5.4.

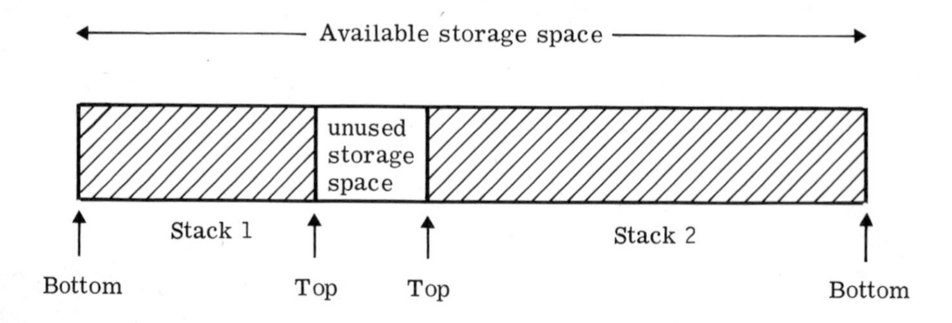


Fig. 5.4. Storage for two stacks

Stack 1 expands to the right into the unused storage space and stack 2 expands to the left into the unused storage space. When the top of stack 1 coincides with the top of stack 2 then we have overflow and all the available storage space has been used.

This solution for two stacks cannot be extended any further since it is impossible to have both a fixed bottom location for the stacks and also to allow overflow to occur only when there is no empty space anywhere in the available storage area. The answer to this problem is to allow for a variable bottom location for each stack and with this change reasonably efficient algorithms due to Knuth and Garwick exist for manipulating n stacks.

### 5.3.4 Knuth's Algorithm for the Manipulation of n Stacks

This is a very straightforward algorithm which uses pointers corresponding to the top and bottom of each stack. The top pointer points to the top item in the stack whilst the bottom pointer points to the location just before the bottom element of the stack. If we try to remove an item from a stack whose top and bottom pointers are equal then underflow occurs because this stack is empty. If we try to add an item to the ith stack when its top pointer is equal to the bottom pointer of the (i+1)st stack then local overflow of the ith stack occurs. The algorithm then searches to see if there is any space available in any of the other stacks. The general picture of the ith stack is given in fig. 5.5.

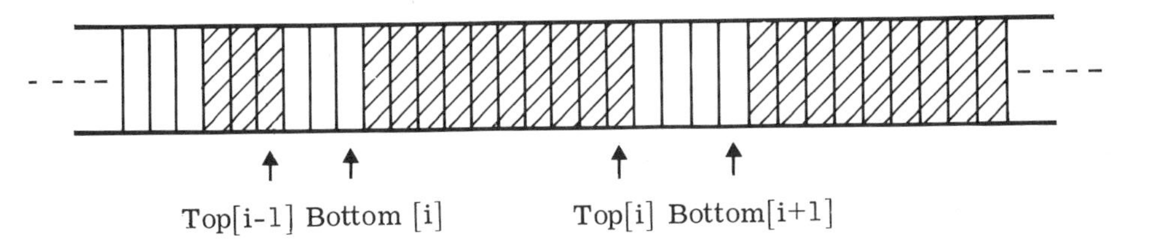


Fig. 5.5. A pictorial representation of ith stack

Knuth's algorithm for searching for storage space when stack i overflows is in three parts:

Step 1. Search to the right until we find the first stack j(> i) which is not using all its storage space i.e. Top[j] < Bottom[j + 1]. Move all the items in stacks j to i + 1 and their top and bottom pointers up one place. In order not to destroy information this must be done by first moving the top item in the jth stack and working down to the bottom item in the (i+1)st stack.

Step 2. If our search in Step 1 reaches the top of the available storage we return to stack i and search to the left until we find the first stack k (< i) which is not using all its storage space, i.e. Top [k] < Bottom [k + 1]. Move all the items in stacks k + 1 to i and their associated top and bottom pointers down one place. In order not to destroy information this must be done by first moving the bottom item of the (k+1)st stack and working up to the top item in the ith stack.

Step 3. If in our search in Step 2 we reach the bottom of the available storage then there is no room in any of the $n$ stacks.

The basic algorithm can be improved in the following ways:

(i) Use a good initial allocation of storage space to the $n$ stacks. For example if each stack is expected to be about the same size then the available storage space is as nearly equally divided between the $n$ stacks as possible. In other cases a different initial allocation of storage may be preferable. However whatever the initial allocation its effect is only noticeable in the early stages.

(ii) The shifting of the items in the stacks when overflow occurs is a time-consuming operation and it would be considerably better if we could move not just one place but several places when the storage has to be repacked. This is the basis of an algorithm due to Garwick which when stack overflow occurs repacks the $n$ stacks according to the changes that have occurred since the last repacking. Naturally this is a good deal more complicated than Knuth's algorithm for $n$ stacks. Extra storage space is required to hold the state of the $n$ stacks when the last repacking occurred. In addition repacking is itself quite a complicated process. There is however one other drawback to such a sophisticated algorithm and that is that the algorithm will perform a large number of elaborate operations as the available storage space becomes nearly full, since quite a high proportion of the insertions could cause repacking. In fact in many such cases the process is about to overrun the available storage anyway and in an algorithm such as Garwick's there would be a great deal of wasted time between the storage being 90 per cent full and completely overflowing. One suggestion by Knuth for avoiding such difficulties is to be resigned in advance to being able to use only part of the total available storage space and to stop the algorithm when the storage is nearly full.

## 5.4 LINKED ALLOCATION OF STORAGE

The sequential allocation of storage for linear lists given in the previous section has many advantages but there are some applications for which such a method of storage is very inflexible. We have seen that overflow presents difficulties with the sequentially stored lists we have considered; all the storage to be used must be consecutive and sets of locations in separate pockets cannot be exploited. An alternative method of storage is to have a link in every node which contains the address of the next node in the list. The last node in the list therefore has a special link (usually denoted $\Lambda$) to indicate that there are no more nodes in the list. This is often called the 'null link' and can be any value which cannot be confused with a possible address. If for example all addresses are positive integers the null link can be a negative integer. Each node therefore can be considered divided into two fields i.e. an Information or Data field and a Link field.

Node X:

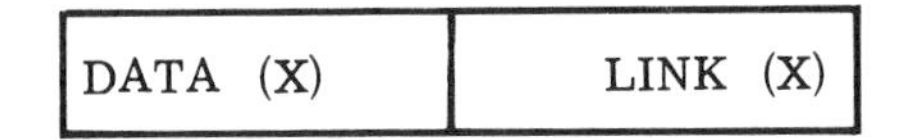


Diagramatically we can draw a linked list of the four nodes, A, B, C and D as follows (fig. 5.6).

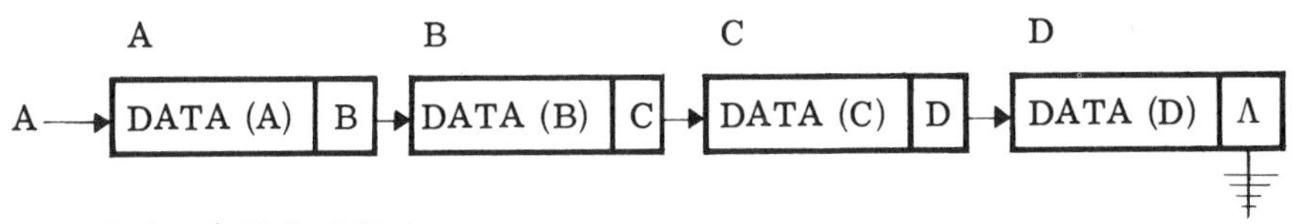


Fig. 5.6. A linked list

A, B, C and D are any four locations in store not necessarily sequential but with addresses given by the links, e.g. LINK (B) = C. Thus in general for a linked linear list we need a pointer to the first node in the list (A in the above example). We can access any other nodes in the list via the first node (often called the list head) and the links.

In order to manipulate a linked allocation of storage system we need a general pool of storage from which we can get nodes and onto which nodes can be returned when they are no longer required. We will consider this general pool of storage to be a linked stack S: see fig. 5.7.

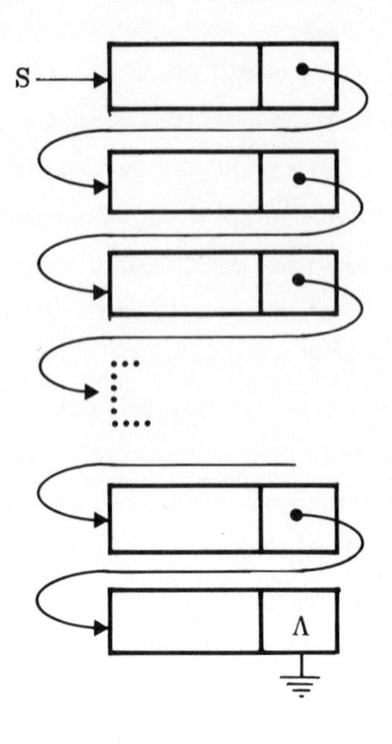


Fig. 5.7. The general pool of storage

The two primitive operations for this general pool of storage are given by the following two procedures.

**procedure** UNSTACK (T, S, POOL EMPTY);
    **pointer** T, S; **label** POOL EMPTY;
    **comment** this procedure performs the primitive operation of getting a node from the general pool of storage and assigning it to the link variable T. In order to do this the top of the stack S is removed and T is made to point to the node removed;
    **begin if** S = Λ **then goto** POOL EMPTY;
                T:= S; S:= LINK(S)
    **end**;

```
procedure STACK (T, S);
        pointer T, S;
        comment this procedure performs the primitive operation of
        returning an unwanted node T to the general pool of storage.
        Thus the node pointed to by T is put on the top of the stack S;
        begin LINK (T):= S; S:= T
        end;
```

High level languages which include linked storage structures (e.g. RECORDS and REFERENCES in ALGOL W) generally do their own storage management, thus relieving the programmer of finding storage space for new nodes and garbage collection.

Generally insertion and deletion from a linear list are considerably easier when the storage is linked rather than sequential. In fact we can see this diagramatically (figs. 5.8 and 5.9). The new links are shown by broken arrows.

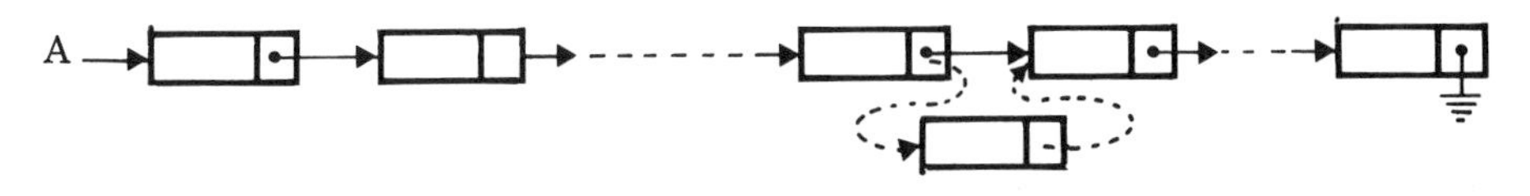


Fig. 5.8. Insertion into a linked linear list

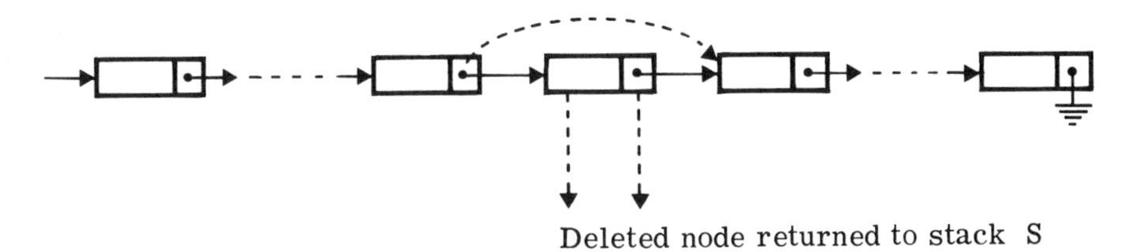


Fig. 5.9. Deletion from a linked linear list

When insertions and deletions take place only at the ends of the linked list we are back to considering stacks, queues, deques etc. However as has been illustrated above the attraction of such special linear lists is less when we use linked allocation of storage since insertion and deletion can in fact be done with not too much difficulty anywhere in the list.

### 5.4.1 Stack Operations

Consider a linked stack which has a stack pointer A. As in section 5.3.1 the only two operations are to load a new element onto the top of the stack and to remove the top element from the stack. The two procedures, given in the previous section for manipulating the general pool of storage will be used in these operations.

(i) To load the element y onto the stack:

UNSTACK(T, S, POOL EMPTY);
**comment** the node T has been removed from the general pool of storage;
DATA(T):= y; LINK(T):= A; A:= T;

(ii) To remove the top element of the stack, put the information into y, and return the node to general pool of storage:

**if** A = Λ **then goto** UNDERFLOW;
T:= A; A:= LINK(T);
y:= DATA(T); STACK(T, S);

These two operations on a linked stack are shown diagramatically in fig. 5.10. These operations are essentially a cyclic permutation of the three links:

| | |
|---|---|
| T | the link to the node to be deleted or inserted, |
| LINK(T) | the contents of the link field of the node to be deleted or inserted, |
| A | the stack link. |

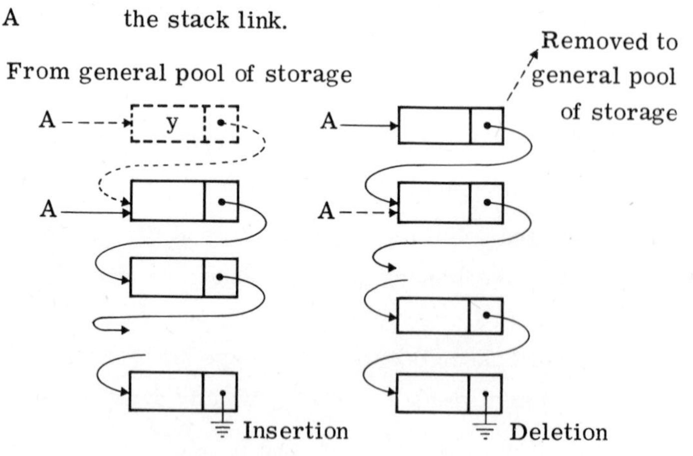


Fig. 5.10. The stack operations for a linked list

## 5.4.2 Queue and Deque Operations

The simplest way to apply linked allocation to queues is to consider the links to run from the front of the queue to the rear. In this manner we can make deletion, which takes place at the front of the queue only, a simple operation since the new front node is directly found. As with sequential allocation of queues we will use two pointers F and R to the front and back of the queue respectively, e.g. fig. 5.11.

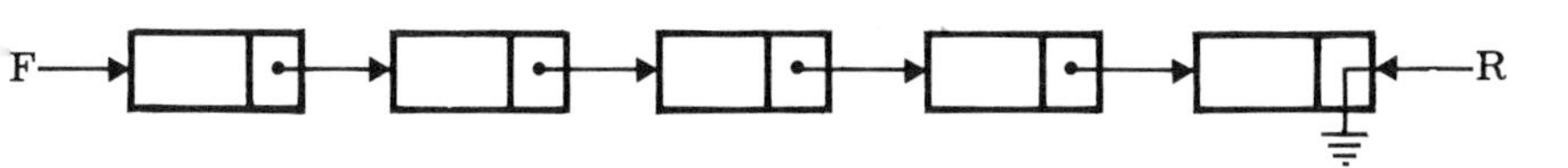


Fig. 5.11. A linked queue

The main difficulty in these simple queue operations is in making sure they work correctly for empty queues. In order to keep the conditions for an empty queue reasonably consistent we will represent it by $F = R = \Lambda$. (Knuth has a more ingenious method of representing empty queues which gives somewhat simpler algorithms at the expense of clarity.) The algorithms for the two queue operations are

i) Load the element y onto the rear of the queue:

UNSTACK (T, S, POOL EMPTY);

**comment** the node T has been removed from the general pool of storage and will now be added to the queue;

DATA (T):= y; LINK (T):= $\Lambda$;

**if** R = $\Lambda$ **then** F:= T **else** LINK(R):= T ;

**comment** the **then** part of this conditional statement is to make sure the front pointer is correctly set when we are inserting into an empty queue;

R:= T;

This insertion operation applied to the queue shown in the previous diagram (fig. 5.11) gives the resulting queue of fig. 5.12.

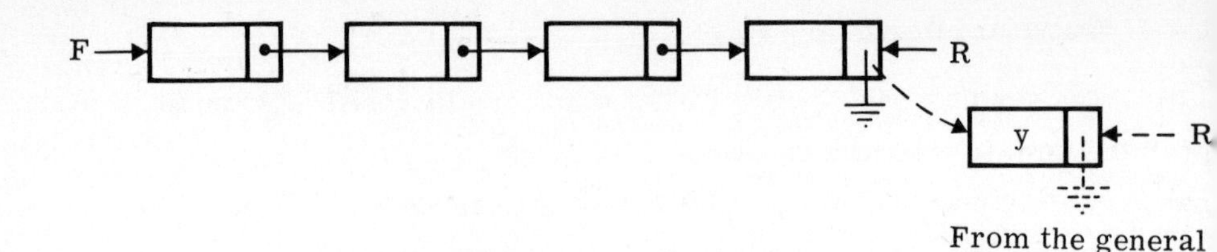


Fig. 5.12. Inserting in a linked queue

(ii) Remove the front of the queue, put the information into y, and return the node to the general pool of storage:

```
if  F = Λ  then goto  UNDERFLOW;

        T:= F;  F:= LINK(T);
        y:= DATA(T);  STACK(T, S);
        if  F = Λ  then  R:=Λ;
        comment  when the last node is re-
        moved from a queue the empty queue
        condition of the rear pointer  R  must
        be set;
```

This deletion operation applied to the queue shown at the beginning of this section (fig. 5.11) gives the queue of fig. 5.13.

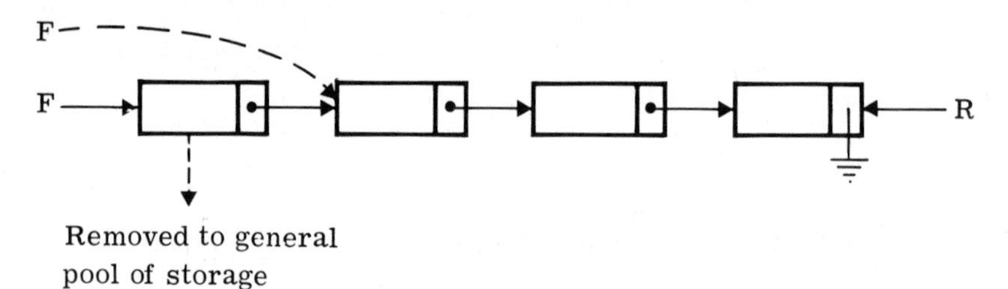


Fig. 5.13. Deletion from a linked queue

A deque with linked storage will be represented in the same way as a queue with front and rear pointers F and R, and the empty deque as F = R = Λ. The two operations (i) and (ii) just given for a queue will also hold for a deque. The other two operations for a deque are

(iii) Load the element y at the front of the deque. (This has been left as an example for the reader.)

This operation is very similar to inserting at the rear of a queue, and the original queue (fig. 5.11) would be altered as shown in fig. 5.14.

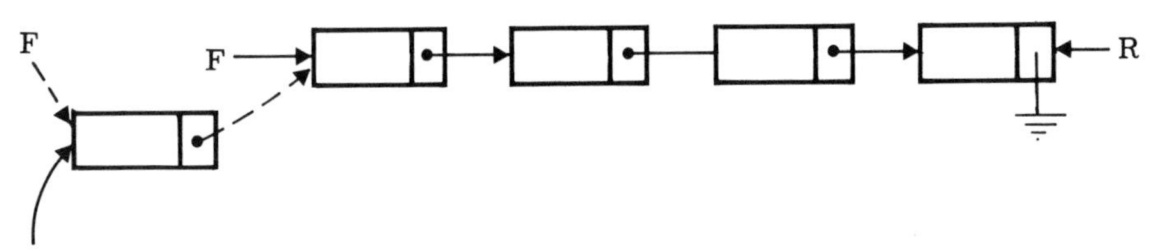


Fig. 5.14. Inserting at the front of a linked deque

(iv) Remove the rear of the deque, put the information into y, and return the node to the general pool of storage.

This is a much more complicated operation than the other three operations on a deque. This is because we have to find the node next to the rear node in order to set its link to $\Lambda$ and to make the rear pointer point to it. The only way to get to this penultimate node with this organization of storage is to work through from the front of the deque until we find it.

```
if F = Λ then goto UNDERFLOW;
if R = F then T:= F:= Λ
         else begin T:= F;
              comment the next statement searches down the
              list until it finds the penultimate node;
                   while LINK(T) ≠ R do T:= LINK(T);
                   LINK(T):= Λ;
         end;

y:= DATA(R); STACK(R, S); R:= T;
```

In this algorithm we have not only to consider the empty deque as a special case but also the deque with only one node. This deletion operation applied to the original queue (fig. 5.11) gives the result of fig. 5.15.

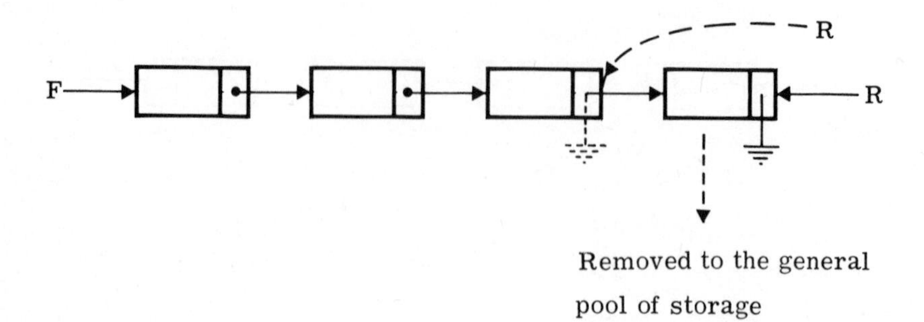


Fig. 5.15. Deletion from the rear of a linked deque.

So far in section 5.4 we have considered a very simple linkage scheme which has a beginning and an end, and we can start at the first node and work to the rearmost node, using the single link that each node has to the next node. Whilst this is quite useful for illustrative purposes it will not necessarily be the best method of linkage in practice. Some of the operations on the structures we have considered are a little awkward at an end and in other applications we may want to traverse the linkage in both directions. Two alternative linkage schemes are considered in the next two sections and they will help to show some of the advantages and disadvantages of other linkage methods.

### 5.4.3 **Circularly Linked Lists** (or Ring Structures)

The simple linked list of the previous sections is altered so that the link of the rearmost node, instead of being the null link, links back the first node. The resulting list is shown in fig. 5.16.

There is no longer any need to think of a first and last node and thus to have the two pointers F and R. We do require one pointer to

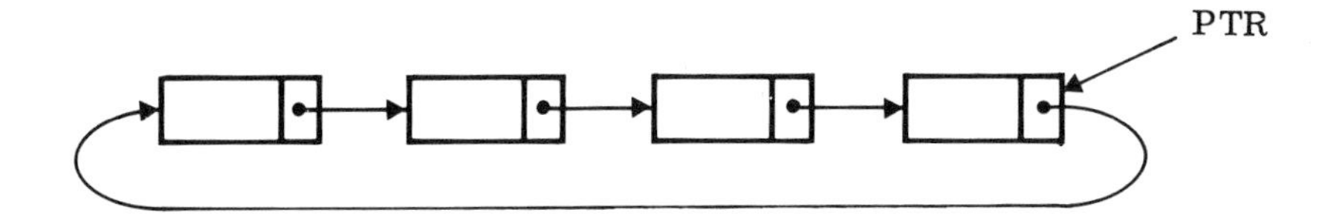


Fig. 5.16. A circularly linked list

the list and in the above diagram we have used the link variable PTR to point to the rightmost node of the list. LINK (PTR) therefore points to the leftmost node of the list. The basic operations for this list assuming the empty list is represented by PTR = Λ are

(i) to insert the information y at the left

```
UNSTACK (T, S, POOL EMPTY); DATA (T):= y;
if PTR = Λ then PTR:= LINK(T):= T
          else begin comment general case of inserting in
                 a non empty list;
                      LINK(T):= LINK(PTR);
                          LINK(PTR):= T;
               end;
```

(ii) to insert the information y at the right. This is the same as operation (i) with PTR:= T; inserted just before the **end**.

In effect we insert at the left and then move the list pointer PTR.

(iii) to set y equal to the information in the leftmost node and delete that node from the list.

```
if PTR = Λ then goto UNDERFLOW;
                  T:= LINK(PTR); y:= DATA(T);
                LINK(PTR):= LINK(T); STACK(T, S);


if PTR = T then PTR:= Λ;
```

This final statement sets the pointer PTR = Λ when we delete the only node in a list of one node and are thus left with an empty list.

(iv) Set y equal to the information in the rightmost node and delete that node from the list (see example [5.9]).

These operations are very similar to those given in the previous section when the linked linear lists used two pointers F and R. We will now examine the use of circularly linked lists to carry out the last three basic operations given in section 4.1.2 namely copying, combining and separating. For example, combining two circularly linked lists L1 and L2 with pointers PTR1 and PTR2 to obtain a combined list with a pointer PTR is performed by the sequence:

```
if PTR2 = Λ then PTR:= PTR1
             else begin if PTR1 ≠ Λ then
                  begin S:= LINK(PTR1);
                        LINK(PTR1):= LINK(PTR2);
                        LINK(PTR2):= S;
                  end;
                        PTR:= PTR2;
                  end;
```

The list L2 is put to the right of the list L1 and there are special cases when either list is empty.

The idea of the leftmost and rightmost node is unnecessary in a circularly linked list which has a natural circular symmetry. In fact in examining such a list we can start anywhere and stop when we return to the starting node. In some applications it is convenient to have a specially recognizable node, usually called the list head, which can be used as a starting and stopping place, such a list is never empty.

### 5.4.4 Doubly Linked Lists

If each node of a linear list has two links, pointing to the nodes on either side of it, then a more flexible method of handling lists is obtained at the expense of extra storage space for the links. Such a doubly linked list can be represented as in fig. 5.17. Each node has two links which will be denoted LLINK and RLINK (i.e. the LLINK points to the node

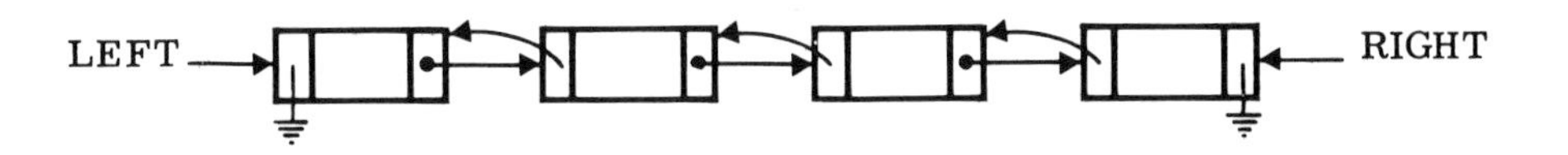


Fig. 5.17. A doubly linked list

to the left) and the whole list has the pointers LEFT and RIGHT at each end. The empty list is represented by LEFT = RIGHT = Λ. For such a list a typical basic operation, for example inserting the information y at the left, is

UNSTACK (T, S, POOL EMPTY); DATA(T):= y;
LLINK(T):= Λ; RLINK(T):= LEFT;
**if** LEFT ≠ Λ **then** LLINK(LEFT):= T **else** RIGHT:= T;
LEFT:= T;

However, a doubly linked list is usually easier to manipulate if we have a specially marked node called the list head and the left link of the leftmost node points to the rightmost node, and the right link of the rightmost node points to the leftmost node: see fig. 5.18.

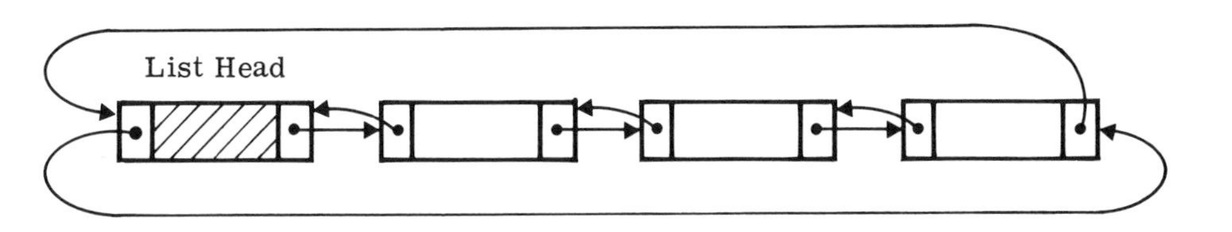


Fig. 5.18. A doubly linked list with a list head

The empty list is now represented by the LLINK and the RLINK of the list head pointing to the list head itself. Such a doubly linked list is symmetrical so that for a general node X

RLINK(LLINK(X)) = LLINK(RLINK(X)) = X

The extra storage for the second link is in many practical cases compensated for by the following advantages of a doubly linked list:

(a) It is very easy to go in either direction in examining the list. Thus the algorithm (iv) given in section 5.4.2 where the only way to find the penultimate node was to start at the front and work forward could be considerably improved with links going in the other direction.

(b) A node X can be deleted given only the value of X. The algorithm for doing such a deletion is

```
RLINK(LLINK(X)):= RLINK(X);
LLINK(RLINK(X)):= LLINK(X);
STACK(X, S);
```

Thus doubly linked lists are ideal when random nodes have to be removed and this removal can be seen diagramatically in fig. 5.19. This contrasts

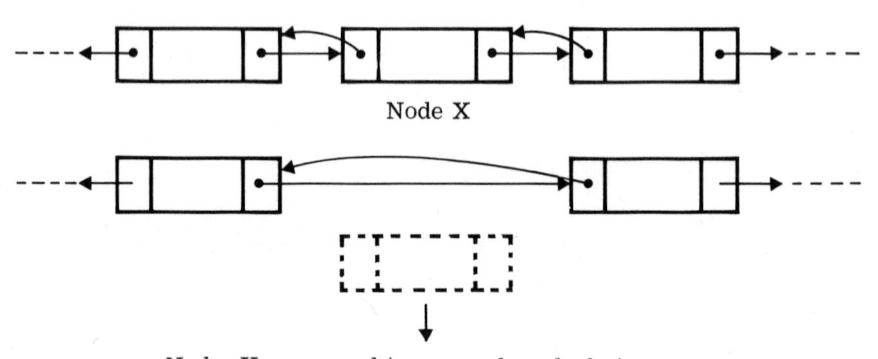


Fig. 5.19. Deleting node X from a doubly linked list

with a singly linked list when we needed to know the preceding node before we could delete node X.

(c) Insertion next to a node X can also be done knowing only the value of X. The algorithm for inserting the information y to the left of an existing node X in a doubly linked list is

```
UNSTACK(T, S, POOL EMPTY); DATA(T):= y;
RLINK(T):= X; LLINK(T):= LLINK(X);
RLINK(LLINK(X)):= T; LLINK(X):= T;
```

This is shown diagramatically in fig. 5. 20.

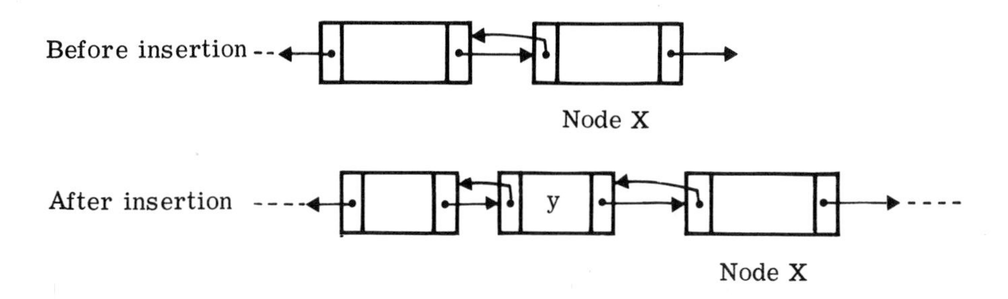


Fig. 5. 20. Insert new node to left of node X in a doubly linked list

There are of course more links to alter when operating with a doubly linked list. In the above algorithm we alter five links as opposed to only three links in a singly linked list.

## 5. 5 COMPARISON OF SEQUENTIAL AND LINKED ALLOCATION OF STORAGE

In this last section on linear lists we intend to make some fairly general remarks about the advantages and disadvantages of sequential and linked storage allocation. However, many of these general remarks may be invalidated in a particular application by the types of data structure available in the programming language or the efficiency of the compiler in implementing various features of the language. Unfortunately, these local disadvantages can quite overshadow the general theoretical gain from using the theoretically most appropriate data structure. Another important feature in deciding between linked and sequential storage will be one which appears frequently in choices of representation and in programming, the need to balance conflicting requirements. In many cases our choice of storage allocation will be a compromise because in one part of the program the operations we have to perform may be better suited to linked storage while in another section of the program sequential storage may be more appropriate and efficient. We have to balance the disadvantage of using the 'wrong' method of storage in a particular part of the program with the frequency that this part of the program is used. Such compromise decisions are often very difficult to make a priori and are often the essence of the art of computer programming. We will now look at the two methods of storage from various viewpoints.

(1) **Amount of Storage.** In general linked storage is worse since each node requires extra space for the link (or links). However, this extra space may be more apparent than real because of other factors. Some possible reasons for smaller total space requirements are

(a) there may be unused storage space already in each node which can be utilized for the link or links.

(b) the same information may be common to several lists and using linked allocation we can use shared storage space. If in fact there is a lot of shared information then linked allocation of storage can use less space overall than sequential storage.

(c) As we saw in section 5.3.3 some algorithms using sequential storage can become very inefficient when they approach a high density loading. Thus in sequential storage we may have to leave empty locations in order to keep a reasonable level of efficiency.

(2) **Insertion and Deletion.** These operations are much easier with linked lists where all we have to do is alter the appropriate linkage. With sequential allocation of storage, insertion is particularly difficult as it involves shifting a large number of elements. Even deletion generally implies shifting elements if we are to take advantage of the deleted storage.

(3) **Accessing Elements.** The comparison between the storage strategies depends on whether we make random references to the items in the list, or whether we work sequentially through the nodes. If we require random access then sequential storage is much more efficient, since we can immediately access the kth item in a sequential list whilst in a linked list we have to work our way from the list head through the links to the kth item. If we access the items sequentially there is not a

great deal of difference between linked and sequential storage although even in this case sequential storage can be rather faster because of the effects of machine architecture upon the construction of addresses.

(4) **Combining and Separating.** These operations are easier to do with a linked list since they only require small changes in the linkage. With sequential storage these operations may require considerable reorganisation of storage and shifting of elements.

## 5.6 BIBLIOGRAPHY

The most important source for extensions of the material in this chapter is Knuth, vol. 1 [1]. In addition to the theoretical material on linked lists Knuth gives three large examples: (i) topological sorting using ordinary linked lists (ii) arithmetic on polynomials to illustrate the use of circularly linked lists and (iii) doubly linked lists applied to simulation of a lift system. A general discussion of information structures has been given by D'Imperio [2]. This paper makes a clear distinction between 'data structures', which are computer-independent and 'storage structures' which are the way of representing a theoretical data structure in computer storage. The majority of this paper is a detailed analysis of the structures used by twelve languages, some for list processing (e. g. LISP, IPL etc. ) and others for string manipulation (e. g. SNOBOL, COMIT). Hoare [3] gives a general approach to data structures and their incorporation into Algol (which led to the **record** and **reference** facilities of Algol W) and compares them with the data structures facilities of Lisp, Simula and PL/I. Wegner's book [4] refers to information structures in the final chapter.

The four new books mentioned in the previous chapter by Berztiss [5], Elson [6], Shave [7], Stone and Siewiorek [8], have also sections on linked lists. Particularly useful is the book by Shave since it includes detailed programs in ALGOL W which use the RECORD and REFERENCE features of that language. In the other books high level languages are also used; in Berztiss FORTRAN is used where possible, in Stone and Siewiorek the language used is mainly Algol-like.

[1] D. E. Knuth: The Art of Computer Programming, vol. 1: Fundamental Algorithms, pp. 228-95. Addison-Wesley, 1968.

[2] M. E. D'Imperio: Data Structures and their Representation in Storage. Annual Review in Automatic Programming, vol. 5, pp. 1-75, edited by M. I. Halpern and C. J. Shaw. Pergamon Press, 1969.

[3] C. A. R. Hoare: Record Handling in Programming Languages, pp. 291-348, edited by F. Genuys. Academic Press, 1968.

[4] P. Wegner: Programming Languages, Information Structures and Machine Organisation. McGraw-Hill, 1968.

[5] A. T. Berztiss: Data Structures - Theory and Practice, 2nd edition, Academic Press, 1975.

[6] Mark Elson: Data Structures, Science Research Associates, 1975.

[7] M. J. R. Shave: Data Structures, McGraw-Hill, London, 1975.

[8] H. S. Stone and D. P. Siewiorek: Introduction to Computer Organization and Data Structures: PDP11 edition, McGraw-Hill, New York, 1975.

## EXAMPLES 5

[5.1] Suggest data structures for representing the following card games in a computer:

(a) Clock patience,

(b) Pontoon (Vingt-et-un),

(c) Solitaire patience.

[5.2] If we wished to simulate on a computer the movements of people in (a) a self-service cafeteria, and (b) at a bus stop, what sort of data structures would be most suitable?

[5.3] For each of the three special linear lists (a stack, a queue and a deque) given the six data items A, B, C, D, E and F on the input stream in that order which of the following permutations of the letters is possible on the output stream?:

(a) ABCDEF (b) BDCFEA (c) AEBDCF

(d) DBACEF (e) ABFDEC (f) EBFCDA

[5.4] Two further special linear lists are defined as follows

Scroll: Deletions and access are allowed at both ends but insertions can only be made at one end of the list.

Output restricted deque: Insertions are allowed at both ends but deletions and access can only be made at one end of the list.

Answer the previous example using these two lists instead of a stack, a queue and a deque.

[5.5] Given a deque in which the elements are stored in sequential locations and which has front and rear pointers F and R, write an algorithm for the primitive operation of loading an element y at the front of the deque. The algorithm should be correct when the deque is empty before loading. (See section 5.3.2.)

[5.6] How would you allocate initially a storage space of 100 locations into seven stacks when using Knuth's algorithm (section 5.3.4) if

(a) their expected sizes are equal,

(b) the first and the last stack are likely to be twice as big as any of the others?

[5.7] Assume a linked list as in fig. 5.6 with a pointer FIRST to the first node in the list. Write an algorithm which will reverse the pointers so the links go in the opposite direction and make the pointer to the list (FIRST) point to the last node in the original list. So fig. 5.6 would become fig. 5.21.

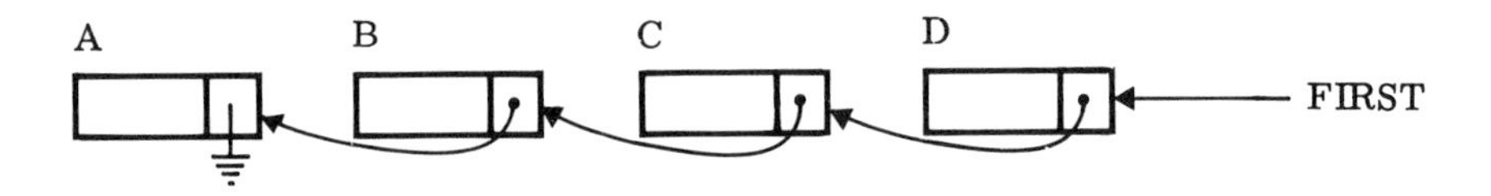


Fig. 5.21. The linked list (fig. 5.6) with the links reversed

[5.8] Write an algorithm for the primitive operation of loading an element y at the front of the deque as in example [5.5] except assume linked storage for the deque elements. (See section 5.4.2.)

[5.9] Given a circularly linked list (fig. 5.16), write an algorithm for the primitive operation of deleting the rightmost node (i.e. the one pointed to by PTR) and setting y to the information field of this node.

[5.10] Write algorithms for a singly linked list (fig. 5.6), a linked queue (fig. 5.11), a circularly linked list (fig. 5.16), a doubly linked list (fig. 5.17) which will carry out each of the following operations and leave the resulting list (if there is one) in the same representation

(i) Joining two lists,

(ii) Splitting a list into two lists,

(iii) Deleting the list and returning its nodes to the general pool of storage.

[5.11] Write an algorithm to insert the information y to the right of the existing node X in a doubly linked list.

[5.12] Define the terms FIFO store and FILO store, and discuss briefly one use for each. Write Algol statements which carry out the actions of (a) adding an item to, and (b) retrieving an item from, ONE such store, represented as an Algol array, stating which type of store you are representing.

(Glasgow 1967)

[5.13] Define a stack and a deque.

Given the permutation 1 2 3 4 5 on the input stream of a stack, which of the permutations,

(i) 1 3 5 4 2

(ii) 5 1 3 2 4

(iii) 3 2 1 5 4

(iv) 2 3 5 1 4

can be obtained on the output stream? What would the result be if the stack was replaced by a deque? When a permutation is obtainable show how to do it.

The circular list shown in fig. 5.16 consists of a (finite) set of nodes each consisting of two fields; an INFORMATION field and a LINK field. Give a representation of a node and use it to write an algorithm which will reverse the directions of the pointers; make sure your algorithm will work for an empty list.

(Newcastle 1969)

[5.14] Explain the purpose of address pointers in each of the following:

(a) list structures;

(b) tree sorting;

(c) sorting with detached keys;

(d) subroutine linkage.

(Leeds 1970)

[5.15] What is a stack, and how may it be implemented in ALGOL?

How would you program a computer to find, and print in descending order, all the distinct prime factors of a given positive integer? Give your algorithm in the form of a flow diagram, with additional explanation as needed; a computer program is <u>not</u> asked for.

(Leeds 1971)

[5.16] Survey the relative merits of sequential and linked linear lists. Write an algorithm which will change the links in a linear linked circular list so that the order of the elements in the list is reversed.

Would you be justified in choosing a method different from the linked circular method for storing a list which needed reversing often? Explain and give reasons for your answer.

(St. Andrews 1971)

[5.17] It is required to hold a fixed number of tables in the core store of a computer. Each table is represented by a set of consecutive words, and provided with an initial allowance of space in which to expand. Describe a system for rearranging the tables in the store whenever one of them is about to overflow the space allocated to it. How should cross references between tables be effected in such a system?

(Essex 1971)

[5.18] The N - 1 jobs in an operating system are stored in M queues depending on their status. Each job is in one, and only one, queue at any one time. The queues are represented by an M + N word table, the M entries with a negative offset with respect to the symbol Q are the queues heads, the N positive entries the jobs. The right half of each entry contains a pointer to the following entry in the queue or if this entry is the last entry in the queue a pointer to the queue head. The left half is similarly used to point to the previous entry in a queue. For example, fig. 5.22.

Write an assembly language routine to move a job x from its present queue to either the front or back of a new queue y. Describe the operation of the routine using flow charts and prose. Describe the format of the parameters which you assume.

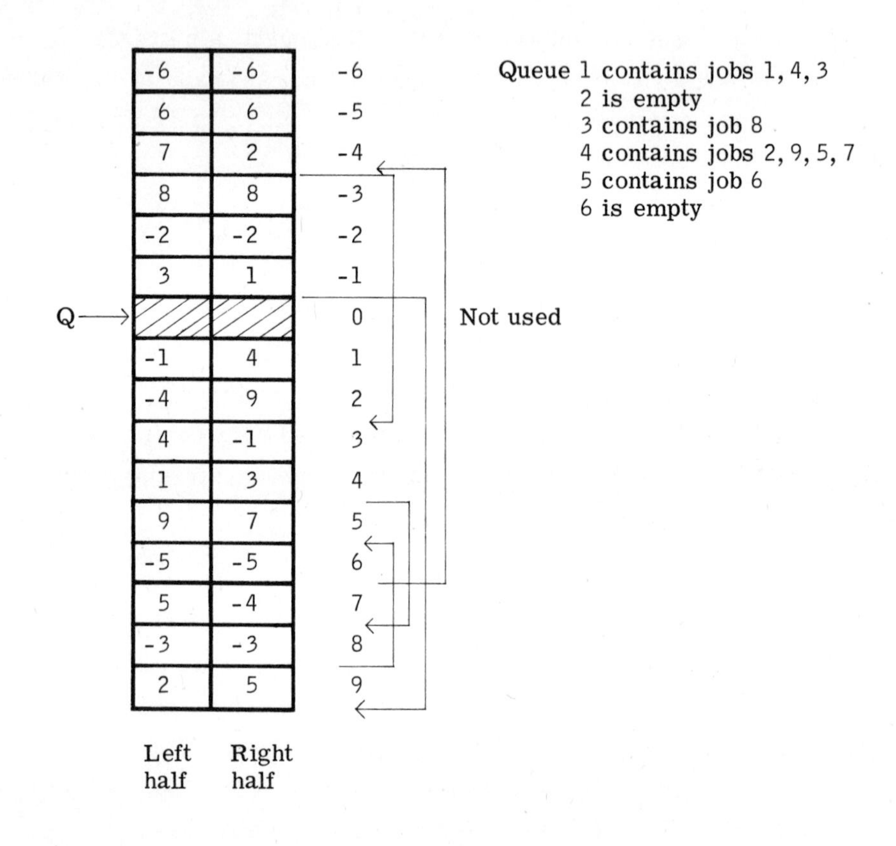


Fig. 5.22

(Essex 1971)

[5.19] It is proposed to use a list as a stack for storing atoms on a last in first out basis. Declare the ALGOL procedure

(a) in (atom, stack) for adding atom to the stack held in list stack and the function designator,

(b) out (stack) which removes the top atom from the stack and returns as its result a pointer to the atom removed.

Write the following procedures to manipulate the top atoms on the stack:

(c) dup (stack) to give two copies of the atom on top of the stack so that (a, b) becomes (a, a, b),

(d) rev (stack) to reverse the order of the two topmost atoms so that (a, b, c) becomes (b, a, c),

(e) perm (stack) to permute the top three atoms so that (a, b, c, d) becomes (b, c, a, d).

How could a list be used to represent a queue on a first in first out basis? Discuss times needed to add or remove items from the queue and the possible effects on such times of using a list structure incorporating both forward and backward pointers.

(Leeds 1971)

[5.20] The linked linear list consists of a finite set of nodes, each consisting of two fields; an INFORMATION field and a LINK field.

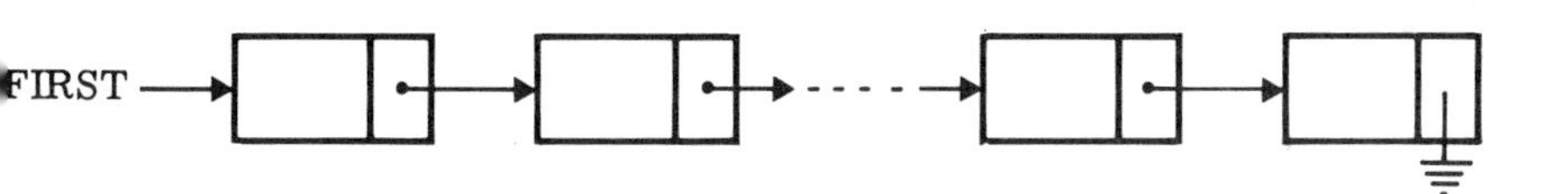


The link to the first element of the list is in FIRST and the last node contains the null link in its LINK field. Write algorithms which will carry out the following operations on these lists.

(a) Join the two lists with links FIRST and SECOND and put the link to the new list in THIRD.

(b) Reverse the directions of the links so that the resulting list is:

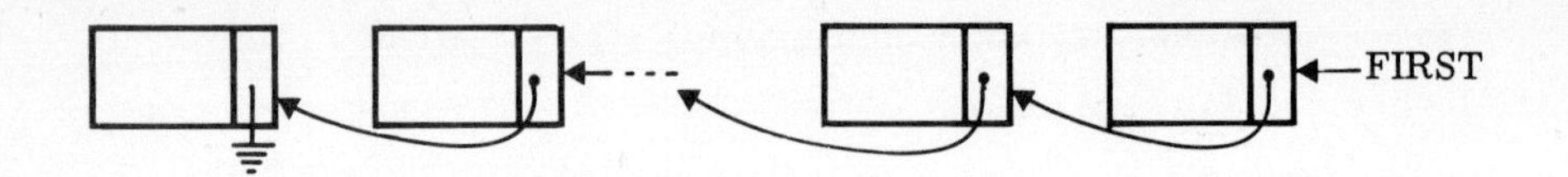


Make sure your algorithms will work for an empty list.

Show how a linked linear list can be used as a stack.

Given the permutation 1, 2, 3, ... n on the input stream of a stack, show that the permutation $p_1, p_2, p_3, \ldots p_n$ cannot be obtained on the output stream of the stack if there exists $p_i < p_j < p_k$ with $k < i < j$. Give an example for $n = 5$.

(Newcastle 1972)

[5.21] Suppose an algebraic polynomial is held as a linked list in which each node has 3 fields -

POWER - the power of x being represented
COEFF - the coefficient of $x^{POWER}$
NEXT - a pointer to the next node

The terms are held in increasing order of power and no term with a zero coefficient appears.

e.g. $3x^2 - 5$ would be held as

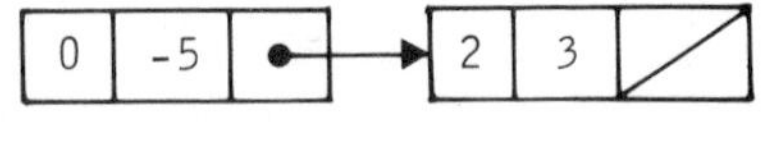


and not as

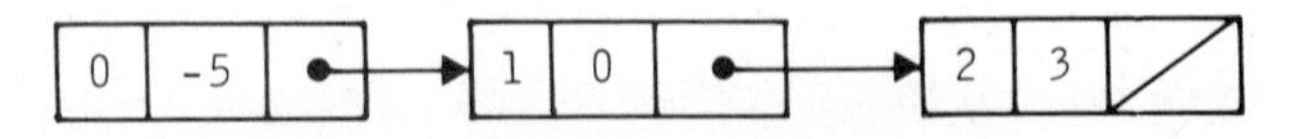


Write an algorithm (preferably using Algol W) which takes 2 parameters which are pointers to such polynomial lists and adds the polynomials together to give a new list, a pointer to which is to be the result. You may find that a recursive algorithm is easier to specify.

(St. Andrews 1975)

[5.22] Programming Project: The Storage Organization for n Stacks

Write a computer program to compare the Knuth and Garwick algorithms (section 5.3.4) for the storage organization for n stacks. In particular compare the algorithms when the available storage is only about 75 per cent full and when it approaches 100 per cent utilization. (For additional detail see Knuth vol. 1, pp. 242-8.)

[5.23] Programming Project: Organizing a Dictionary

It is useful to store the entries in a dictionary in such a way that common prefixes share common storage.

```
C  ( A   ( K   E   -
   (     ( T   -
   ( O     M   ( E   -
   (           ( P   U   T   E   ( -
   (                             ( R   -
   ( R     O     W   -
```

This is best done by arranging some sort of linked structure to which new elements can easily be added. Presumably each entry has some associated information which must be stored (e.g. a list of synonyms). Write a program to read a sequence of records, each with an entry and associated information and to construct a dictionary. Use this to print an alphabetic list of the entries and the associated information. In the special case, when associated information is a list of synonyms, write a procedure which will print all words which are synonymous with a given word.

[5.24] Programming Project: Lift versus Paternoster Project

The purpose of this project is to investigate traffic patterns in an idealized model of a lift and paternoster (an endless vertical moving belt of compartments) in a tower building. This is to be done by writing a program which simulates the traffic flow and then using this simulation program to carry out a small series of experiments.

The program should model a building which has 10 floors and is serviced by one lift and one paternoster. The lift takes $1\frac{1}{2}$ seconds to travel between adjacent floors. If the lift stops at a floor, it remains there with doors open for 10 seconds. The lift gives precedence to requests from floors which it is moving towards. It will change direction in response to a request only when it has reached one end of the lift shaft, or when there are no waiting requests ahead in its current direction. When there are no further requests at all it will remain at its current position. The lift can hold a maximum of 10 people.

The paternoster moves continuously, taking 5 seconds between floors. Cabins arrive at 3 second intervals and each cabin can hold up to two people. It is illegal to travel around the top or bottom of the paternoster loop!

Every t seconds a group of n people arrive at floor i, with the intention of travelling to floor j. The value of t is an input parameter, n is chosen at random (i. e. with equal probability) from the integers 1, 2, 3 and 4. The value of i is chosen at random from the integers 1-10 and j from amongst the remaining 9 integers. The group as a whole will decide to use the lift if they have to travel a distance of f or more floors where f is an input variable (the group will, of course divide to travel if necessary).

The simulation program is to be used to investigate the influence of the values of t and f on d, the average time it takes for a person to get to his destination floor.

The program should be documented adequately for its correctness to be verified and for it to be used by someone other than its author. The results of the experiments should be described in a memorandum, written as though it were addressed to the architect who designed the tower.

[5.25] Programming Project: Variable Length Arithmetic

A decimal integer can be stored as a linear list of its digits. Design procedures to perform the usual arithmetic operations on integers stored in this way and to read and print them. Note that it is probably better to store integers with digits in reverse order. Was 10 the best base to choose? Can your scheme be extended to incorporate real numbers (or alternatively, rational numbers)? How do you cope with the representation of an irrational number in a given base?

# 6 · Information Structures 3: Trees

## 6.1 INTRODUCTION AND BASIC DEFINITIONS

The information structures examined in the previous two chapters have had linear relationships between the nodes. We now want to extend these concepts to non-linear structures and in particular to trees, which are the most important non-linear structures for the computer scientist. Tree structures have a branching or hierarchical relationship between nodes, and the name implies analogies with both family trees and natural trees. In fact much of the notation in tree theory comes from these two sources.

The classical definition of a tree is as a particular form of a linear graph, which is defined as follows:

A linear graph is a collection of nodes (usually called points) together with a collection of relationships (usually called lines) which describe the connections between the points. A linear graph is connected if every pair of points is joined by a path (i.e. a collection of lines of the form $p_1p_2$, $p_2p_3$, ... $p_{k-1}p_k$ with the points $p_1$ and $p_k$ distinct).

A tree can now be defined as a connected linear graph with no closed circuits or loops. From the above definitions we can show that a tree has the following important properties:

(a) Any two nodes in a tree are connected by a unique path,

(b) A tree with $n$ nodes contains $n - 1$ lines.

The most useful type of trees used in computer representation are rooted trees. A rooted tree is defined as a tree in which one node called the root is given special significance.

A typical tree is shown in fig. 6.1 with the root A drawn at the top.

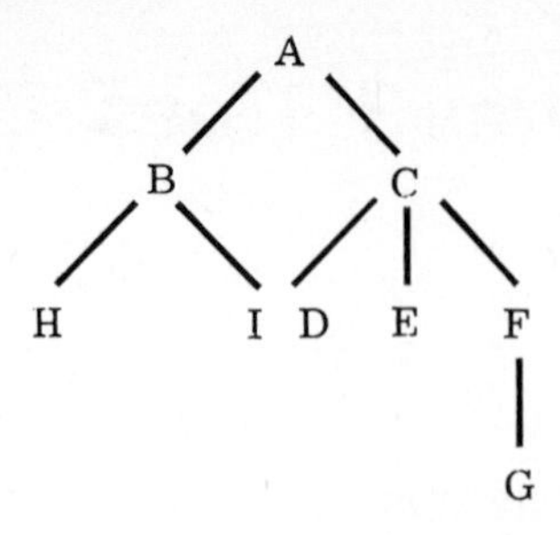


Fig. 6.1. A rooted tree

The other common method of drawing trees places the root at the bottom as in nature. We have adopted the root at the top convention because it always seems more natural and easier to draw trees this way when you are building them up and do not know how much space on the paper you will require for the finished product; it seems, too that this convention is becoming more generally accepted in computer science. The way we draw trees is important in our description of algorithms because the root and the way of drawing give a tree a structural direction so that for example, we can talk about down or away from the root, the rightmost node and higher nodes or working our way back up the tree to the root.

An alternative approach is to define trees recursively; for example

A rooted tree can be defined as a finite set T of one or more nodes such that:

(i) there is one specially designated node called the root of the tree, and

(ii) the remaining nodes (excluding the root) are partitioned into $m \geq 0$ disjoint sets $T_1, T_2, \ldots T_m$ and each of these is in turn a tree. The trees $T_i$ $(i = 1, 2, \ldots m)$ are called the subtrees of the root.

This definition is recursive since it defines trees with n nodes in terms of trees with less than n nodes. One of the important points about recursion is that we must be able to stop the recursion at some level. In this case the definition ends when $n = 1$ and we have a tree which is just a root. Recursion is recognized as one of the fundamental concepts of computer science; a recursive definition for trees is particularly apposite since trees have a natural recursion which can be seen not only in the definitions and their pictorial representations but also in the algorithms for manipulating them.

In drawing a tree as in fig. 6.1 or in representing them in a computer we have added an implicit ordering which was not present in the original definitions. Consider the two trees drawn in fig. 6.2.

Fig. 6.2.

Are these two trees to be considered the same or different? It clearly depends on whether the order of the subtrees is important or not. If it is, then we have <u>ordered trees</u> and in that case the trees in fig. 6.2 are different; if order is not important we have <u>oriented trees</u> and then the two trees in fig. 6.2 are the same. (This difference between ordered and oriented trees is analogous to the difference between permutations and combinations.) As we noted before, when we represent a tree in a computer we necessarily give it an ordering and as computer representation is our purpose, we will only consider ordered trees.

Since trees will be used as a convenient representation of other entities we must be clear what we mean if we say two trees are the same. Two trees are said to be <u>similar</u> if they have the same structure; or more precisely two trees are similar if either they are both empty or all their corresponding subtrees are similar. Two similar trees therefore have the same shape. We can extend this idea to say two trees are equivalent if they are similar and have the same information at corresponding nodes.

Other terms which will be encountered in studying trees are given below:

A forest is a set of disjoint trees.

A free tree is a tree that has no root.

A terminal node or leaf is a node which has no subtrees.

A branch node is any node which is not a leaf. Thus all nodes are divided into two types called branch nodes and terminal nodes.

The degree of a node in a tree is the number of subtrees of that node. (Readers should note that another definition which may be met is the special case of the degree of a node in a linear graph which is the number of lines attached to that node - thus giving one more than the definition used here.)

The level of a node in a tree is 1 if it is the root and is the number of nodes passed through on the path from the root to node A (inclusive of both the root and node A) for any node other than the root.

The length between two nodes is the difference between their levels.

**Example.** In fig. 6.1 the node F is a branch node with degree 1 and is at level 3. The length from node C to node G is 2.

Trees are often spoken of as though they were family (or genealogical) trees. Each root is said to be the father of the roots of its subtrees, which in turn are the sons of the father. The roots of the subtrees of the same father are said to be brothers. The father-son-brother is the basic terminology but it can be extended in the usual way to grandfathers, uncles, cousins etc. Other authors have used slightly different connotations such as mother-daughter-sister or parent-offspring-sibling, the latter of which seems particularly clumsy.

**Example.** In fig. 6.1, if we consider node F the other nodes are related to him as follows. Node A is his grandfather, node B is his uncle, node C is his father, nodes D and E are his brothers, node G is his son and nodes H and I are his cousins.

### 6.1.1 Binary Trees

A binary tree is a tree in which each node has exactly zero or two sons. This is a very important special case of a tree and one which occurs frequently in computing applications.

Unfortunately the definition given above is not universal and Knuth (for example) defines a binary tree recursively as a finite set of nodes which either is empty or consists of a root and two disjoint binary trees (called the left and right subtrees). Obviously the two concepts are similar in many respects but by Knuth's definition a binary tree is not a special case of a rooted tree. In order to avoid confusion in this text the term binary tree will be used as a general term when either definition is applicable. The binary tree as defined by Knuth will be referred to as a Knuth binary tree and that defined at the beginning of this section as a strictly binary tree.

An example of a strictly binary tree and a Knuth binary tree are given in fig. 6.3.

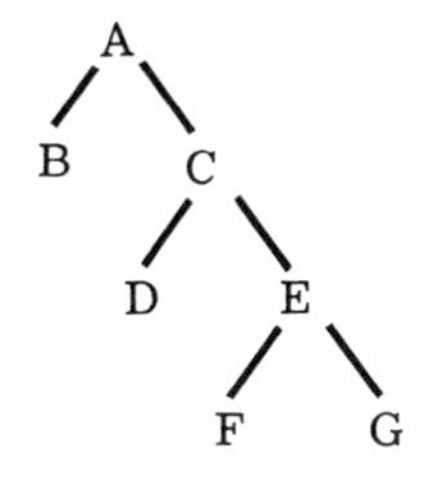


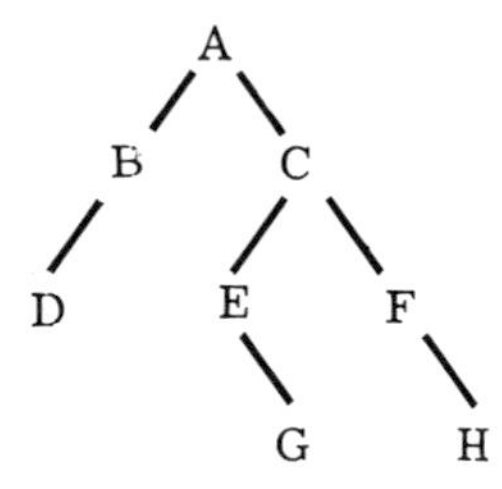


Fig. 6.3. (a) A strictly binary tree (b) A Knuth binary tree

Binary trees figure largely in what follows: they have a natural two-link memory representation and we shall see that all trees can be transformed into binary trees. In many practical applications the resulting tree is in fact a binary tree.

### 6.1.2 Examples of Tree Structures in Practice

**(a) Arrays.** The rectangular array (chapter 4) can be considered as a special case of a tree structure; the **array** X[1:4, 1:2] can be represented as in fig. 6.4.

The first generation nodes correspond to the rows of the array but this representation of a matrix does not show the column relationships between the elements, and from this point of view, the tree does not exhibit all the matrix structure.

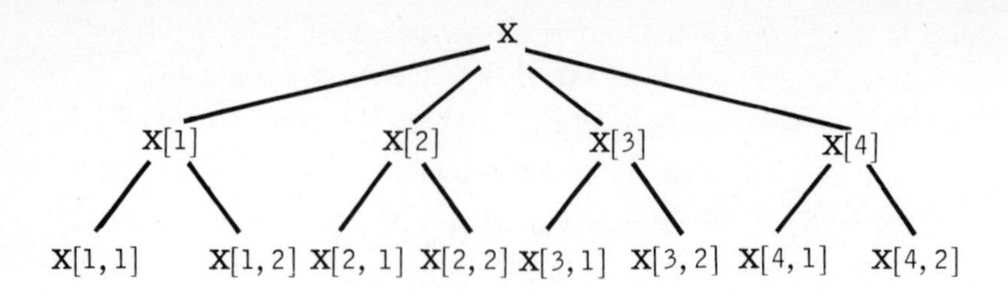


Fig. 6.4. An array as a tree structure

**(b) Arithmetic Expressions.** Arithmetic expressions and Algol statements can be represented as a tree structure as is shown in fig. 6.6. In this representation the operators are the branch nodes and the operands are the leaves. Since the operators used are all diadic, the tree is a strictly binary tree. The rules for evaluating such an arithmetic expression (i. e. the precedence of the operators) are included in the structure of the tree. Such tree structures are useful to compiler writers when setting up the code to evaluate expressions.

**(c) Decision Trees.** This type of tree occurs in such applications as game playing where at each node the branches represent the possible moves (or decisions) that can be made. However even for such a simple game as noughts and crosses the decision tree for the game gets very big indeed because of the large number of different sequences of moves in the game. At level 2 there are 9 nodes since the first player can choose any one of nine squares, at level 3 there are $9 \times 8$ nodes, at level 4 $9 \times 8 \times 7$ nodes and so on.

**(d) Library Classifications.** Several of the library catalogue systems have a tree structure, for example the Dewey Decimal system and its derivatives specify first a broad area of knowledge and successively finer divisions. The numbering in this book is a simple example - chapters and sections only.

There are many other applications of trees in computing; one has already been encountered in chapter 2 illustrating Huffman's coding algorithm. Others occur in searching (chapter 7) and sorting (chapter 8).

## 6.2 TRAVERSING A TREE

The two major problems we are going to examine in this chapter are firstly how to traverse a tree, i.e. methods of visiting each node of the tree once and only once and secondly how a tree may be represented in computer memory. These two problems are, of course, related: the best method of traversing a tree will usually depend on how the tree has been stored. The importance of finding good methods of traversing arises from the frequency with which such operations are required in applications; whenever a node with certain characteristics has to be identified some traversing is undertaken while a comprehensive survey of the data in the structure calls for a complete traverse.

Conceptually one of the simplest ways of traversing a tree is to start at the root, then to traverse all the nodes at level 2 (left to right say), then all the nodes at level 3 (left to right) and so on. This is often known as a level-by-level traversal or a constant depth traversal. The method described above is also called 'top-down' because we start at root and work down the tree. A similar constant depth traversal of a tree could start at the level farthest from the root and work up level-by-level until the root is reached - the 'bottom-up' traversal.

For the tree given in fig. 6.5 the two level-by-level traversals give

| | |
|---|---|
| Top-down | A B C D E F G H I J K L M N |
| Bottom-up | M N I J K L E F G H B C D A |

We can in fact traverse trees in very many different ways and we shall now describe some of the traversals which have some significance or have proved useful in applications.

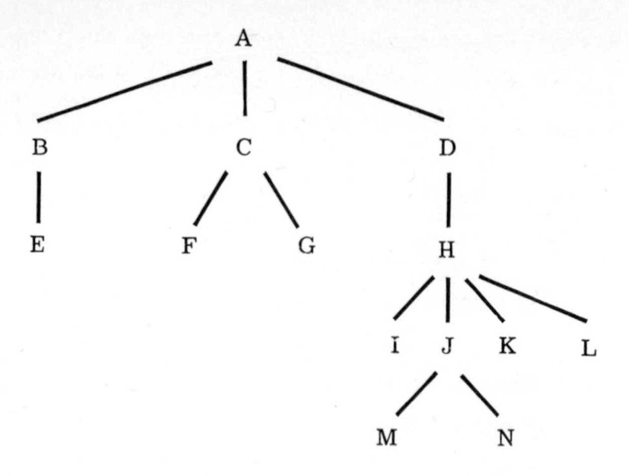


Fig. 6. 5. A typical tree

**Preorder traversal** (Sometimes called the prefix walk over the tree). Start at the root. From each branch node proceed to the leftmost (eldest) son, from each leaf proceed to the right (next youngest) brother or if there is no right brother to the father's brother or, if the father has no right brother, to the grandfather's right brother etc. Stop at the attempt to find the right brother of the root.

The preorder traversal for the tree in figure 6. 5 gives

A B E C F G D H I J M N K L

**Suffix walk.** This traverse proceeds through the terminal nodes starting at the leftmost leaf. The traverse of the terminal nodes is interrupted to traverse a branch node immediately after all the sons of that branch node have been visited. Stop when the root of the tree has been traversed.

The suffix walk for the tree in fig. 6. 5 is

E B F G C I M N J K L H D A

In this type of traversal all the sons are visited before their father.

**Family order traversal.** Tree traversal methods can often be conveniently defined recursively (see Knuth for several such definitions). We will illustrate such a recursive definition of family order traversal, noting first that traversing an empty tree or subtree consists of doing nothing.

1. Visit the root of the first tree.
2. Traverse the remaining trees in family order.
3. Traverse the subtrees of the first tree in family order.

Consider how this definition can be applied to the tree given in fig. 6.5. The steps would be:

|  |  | |
|---|---|---|
| Step 1 | Visit root of tree A | A |
| Step 2 | Traverse the remaining trees in family order | - |
| Step 3 | Traverse the subtrees of tree A in family order | |
| 3.1 | Visit the root of the first subtree of A | B |
| 3.2 | Traverse the remaining subtrees of A in family order | |
| 3.2.1 | Visit the root of tree C | C |
| 3.2.2 | Traverse the remaining trees in family order | |
| 3.2.2.1 | Visit the root of tree D | D |
| 3.2.2.2 | Traverse the remaining trees in family order | - |
| 3.2.2.3 | Traverse the subtrees of tree D in family order | |
| 3.2.2.3.1 | Visit the root of tree H | H |
| 3.2.2.3.2 | Traverse the remaining trees in family order | - |
| 3.2.2.3.3 | Traverse the subtrees of tree H in family order | |
| 3.2.2.3.3.1 | Visit the root of tree I | I |
| 3.2.2.3.3.2 | Traverse the remaining trees in f. o. | |
| 3.2.2.3.3.2.1 | Visit the root of tree J | J |
| 3.2.2.3.3.2.2 | Traverse the remaining trees in f. o. | |
| 3.2.2.3.3.2.2.1 | Visit the root of tree K | K |
| 3.2.2.3.3.2.2.2 | Traverse the remaining trees in f. o. | |
| 3.2.2.3.3.2.2.2.1 | Visit the root of tree L | L |
| 3.2.2.3.3.2.2.2.2 | Traverse the remaining trees in f. o. | - |
| 3.2.2.3.3.2.2.2.3 | Traverse the subtrees in f. o. | - |
| 3.2.2.3.3.2.2.3 | Traverse the subtrees of tree K in f. o. | - |

3.2.2.3.3.2.3 Traverse the subtrees of tree J in f.o.

3.2.2.3.3.2.3.1 Visit the root of tree M — M

3.2.2.3.3.2.3.2 Traverse the remaining trees in f.o.

3.2.2.3.3.2.3.2.1 Visit the root of tree N — N

3.2.2.3.3.2.3.2.2 Traverse the remaining trees in f.o. — -

3.2.2.3.3.2.3.2.3 Traverse the subtrees of tree N in f.o. — -

3.2.2.3.3.2.3.3 Traverse the subtrees of tree M in f.o. — -

3.2.2.3.3.3 Traverse the subtrees of tree I in f.o. — -

3.2.3 Traverse the subtrees of tree C in f.o.

3.2.3.1 Visit the root of tree F — F

3.2.3.2 Traverse the remaining trees in f.o.

3.2.3.2.1 Visit the root of tree G — G

3.2.3.2.2 Traverse the remaining trees in f.o. — -

3.2.3.2.3 Traverse the subtrees of tree G in f.o. — -

3.2.3.3 Traverse the subtrees of tree F in f.o. — -

3.3 Traverse the subtrees of tree B in f.o. — -

3.3.1 Visit the root of tree E — E

3.3.2 Traverse the remaining trees in f.o. — -

3.3.3 Traverse the subtrees of tree E in f.o. — -

The detailed working of the steps is intended to show the depth of the recursion at each stage. The greatest depth is when traversing L or N when there are nine stages.

In family order all the sons of a father appear together starting with the leftmost (eldest) son. The sons appear after their father but not necessarily immediately after their father. In the above example F and G appear after their father C but all of D's family come between C and his sons.

### 6.2.1 Binary Tree Traversal

The methods of tree traversal given in the previous section will now be applied to the special case of a binary tree. The binary tree we will use to illustrate these traversals is that given by the Algol statement

$$A := b \times c + (3.6 - d \times e)/f \uparrow 4. \qquad (6.1)$$

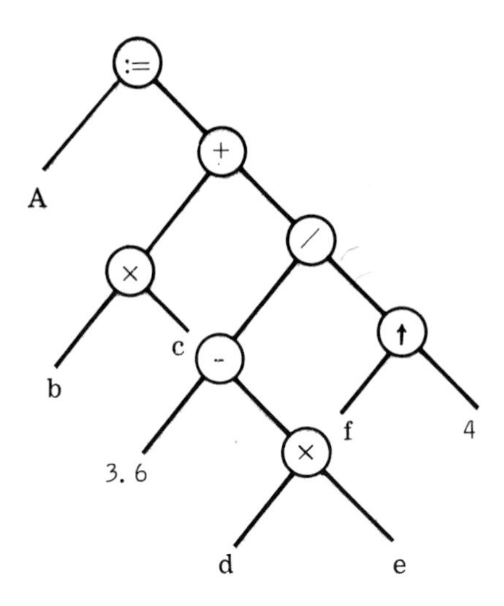


Fig. 6.6. The binary tree for the Algol statement (6.1)

The binary tree for this Algol statement is drawn in fig. 6.6 and the various traversals of section 6.2 give

Constant depth, top down := A + × / b c - ↑ 3.6 × f 4 d e

Constant depth, bottom up d e 3.6 × f 4 b c - ↑ × / A + :=

Prefix walk (preorder traversal)

:= A + × b c / - 3.6 × d e ↑ f 4

Suffix walk

A b c × 3.6 d e × - f 4 ↑ / + :=

Family order traversal

:= A + × / - ↑ f 4 3.6 × d e b c

The two most interesting results are those obtained by the prefix walk and the suffix walk. In the prefix walk the arithmetic expression part of (6. 1) is in what is known as 'Polish' notation while in the suffix walk it is in 'Reverse Polish'. In the first case the operator precedes its two operands and in the latter the operator follows its two operands. Reverse Polish is a very convenient way of writing an arithmetic expression prior to its evaluation (often using a stack) and therefore is of great interest to compiler writers.

None of the methods of traversal give the arithmetic expression in its 'normal' mathematical notation. We can therefore define one which we will call the symmetric order traversal. The recursive definition for this traversal is (do nothing if the tree or subtree is empty)

(1) Traverse the left subtree in symmetric order,

(2) Visit the root,

(3) Traverse the right subtree in symmetric order.

It can easily be verified that if we traverse the tree in fig. 6. 6 by symmetric order traversal we obtain the arithmetic expression in its original 'infix' form. An interesting way of looking at this traversal is to consider the tree structure as a solid wall which we walk round. We start at the northern apex and proceed south always keeping the wall on our immediate left. This can be represented for the Algol statement (6. 1) as shown in fig. 6. 7.

If we output each node as we pass under it we obtain

$$A := b \times c + 3.6 - d \times e / f \uparrow 4 \qquad (6.2)$$

We can also add to this that if the operator at any branch node has lower priority than its father's, or if it has equal priority and the node is a right son, then we output a left bracket when we pass it on the west side and a right bracket when we pass it on the east side. In fig. 6. 7 the - operator has lower priority than its father / and so we have the left bracket before 3. 6 and the right bracket after e and thus obtain the Algol expression (6. 1).

We will discuss tree traversals again when we come to methods of representing trees in section 6. 4.

## 6. 3 THE TRANSFORMATION OF TREES INTO BINARY TREES

As binary trees occupy a very special place in the theory of trees we want to investigate methods of transforming general trees into binary

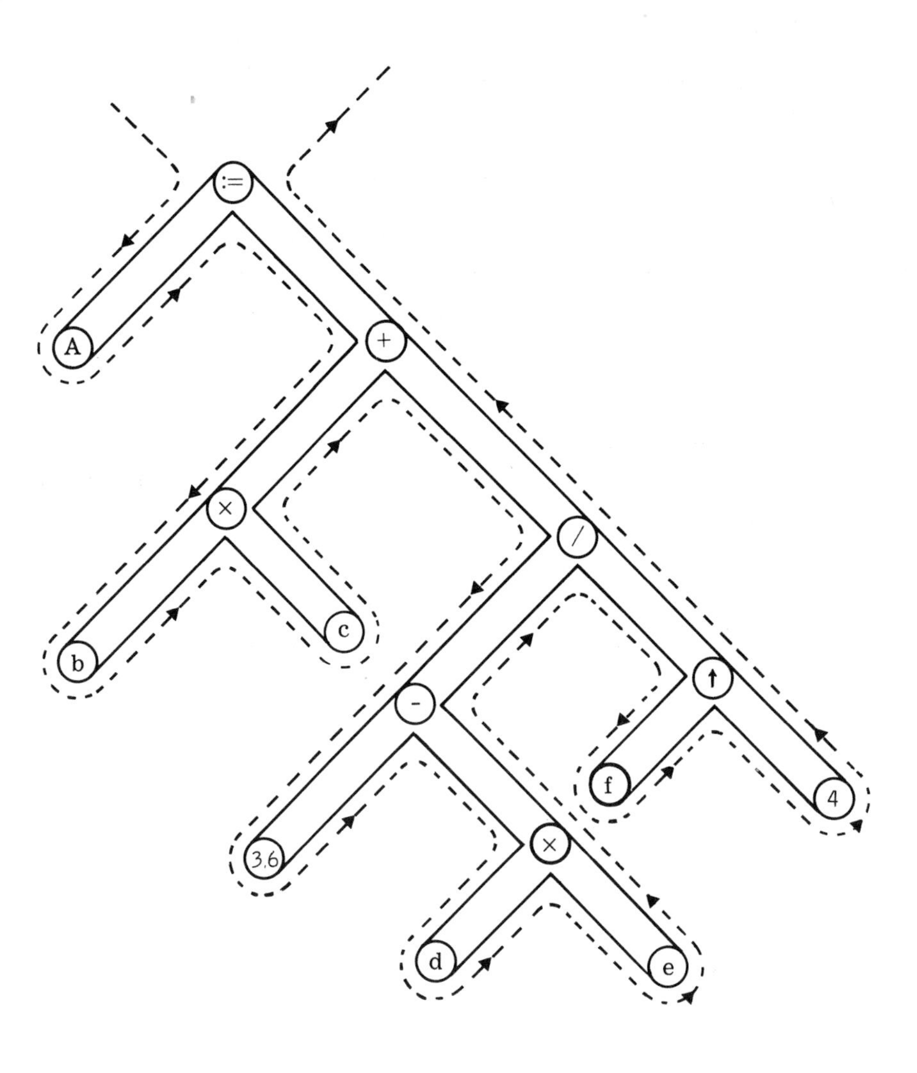


Fig. 6.7. The binary tree for the Algol expression (6.1) represented as a wall with the walk round it

trees. An important transformation is that given by Knuth which we will now discuss in more detail.

### 6.3.1 The Knuth Binary Tree Representation of a Forest

A forest was defined as a set of disjoint trees, and a Knuth binary tree as a finite set of nodes which either is empty or consists of a root and two disjoint binary trees (called the left and right subtrees). A forest can be transformed into a Knuth binary tree by

(i) Linking together the roots of the trees of the forest.

(ii) Linking together all brothers.

(iii) Removing all links from a father to his sons except that from the father to his eldest son.

(iv) Making the root of first tree of the forest the root of resulting Knuth binary tree and tilting the diagram 45°.

An example of this transformation is given in fig. 6.8.

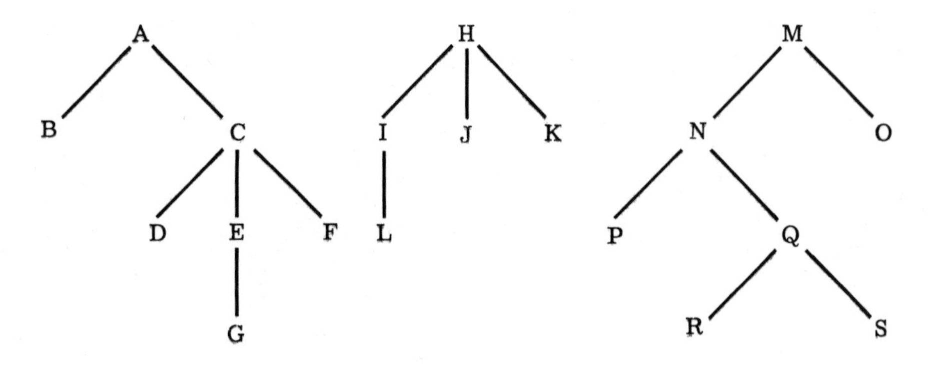


(a) The original three trees

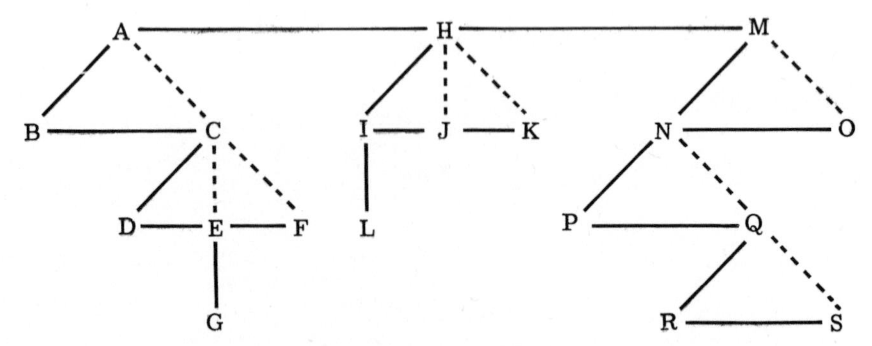


(b) The extra links are added between brothers. The links to be removed are shown with dotted lines

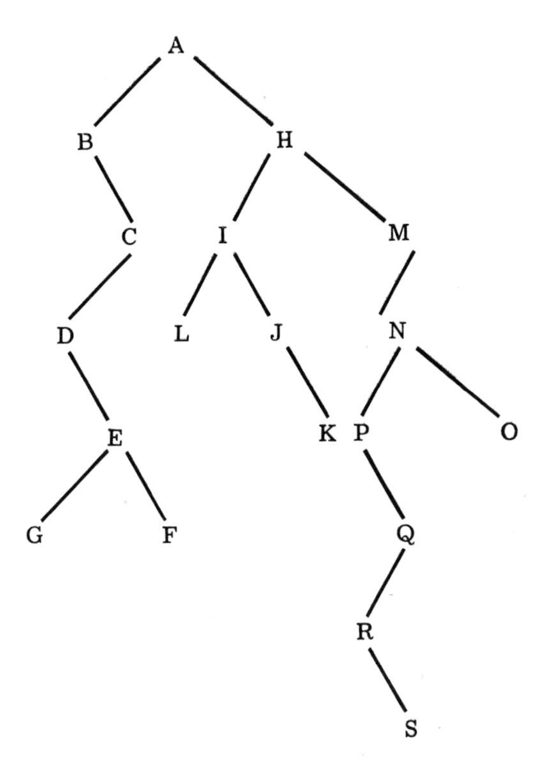


(c) The resulting Knuth binary tree

Fig. 6. 8. The transformation of a forest of three trees into a Knuth binary tree

This transformation can fairly obviously be reversed so that each Knuth binary tree corresponds to a unique forest. This transformation is called a natural correspondence between forests and Knuth binary trees.

An important special case of this transformation is when the forest consists of just one tree. The transformation then gives a Knuth binary tree in which the root has no right subtree.

It should also be noticed that the left subtrees of the resulting Knuth binary tree correspond to links from a father to his eldest son in the original tree. Similarly the right subtrees in the resulting Knuth binary tree correspond to the links added between brothers.

### 6.3.2 Transformation of a Tree into a Strictly Binary Tree

In this transformation the nodes of the original tree are transformed into the terminal nodes of the resulting strictly binary tree. Thus if the original tree has $n$ nodes the resulting strictly binary tree has $(2n-1)$ nodes.

Given an ordered rooted tree we construct the strictly binary tree as follows:

(1) If the tree is a single node, the strictly binary tree is just the root.

(2) If the tree is not a single node, cut the branch between the root and his eldest son. This divides the original tree into two parts which become the left and right subtrees of the root of the strictly binary tree. The left subtree is the part of the original tree rooted on the eldest son of the root and the right subtree is the remainder of the tree (including the root).

(3) Recursively repeat steps (1) and (2) to the two parts of the original tree.

An example of this transformation starting with a 15-node tree is given in fig. 6.9. The first cut is between nodes A and B and this gives the second diagram, fig. 6.9(b), where the left and right subtrees of the root of the strictly binary tree are shown in dotted circles. The cutting operation is now applied recursively and in this way we gradually break down the tree rooted at B. Fig. 6.9(c) shows this process partly completed. Then we return to that part of the original tree rooted at A which is the right subtree in fig. 6.9(b). Again the cutting process breaks

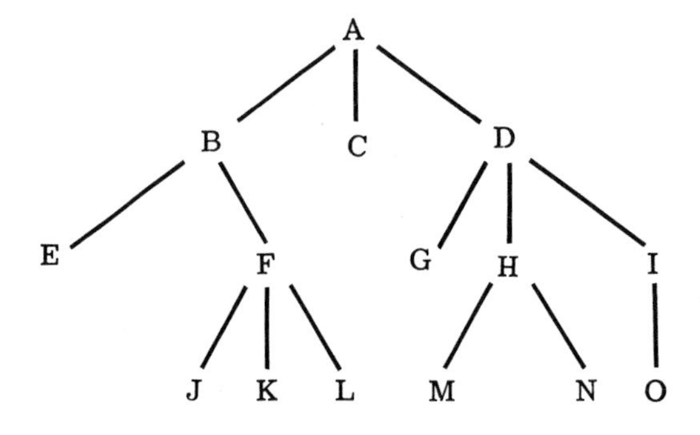


(a) The original tree

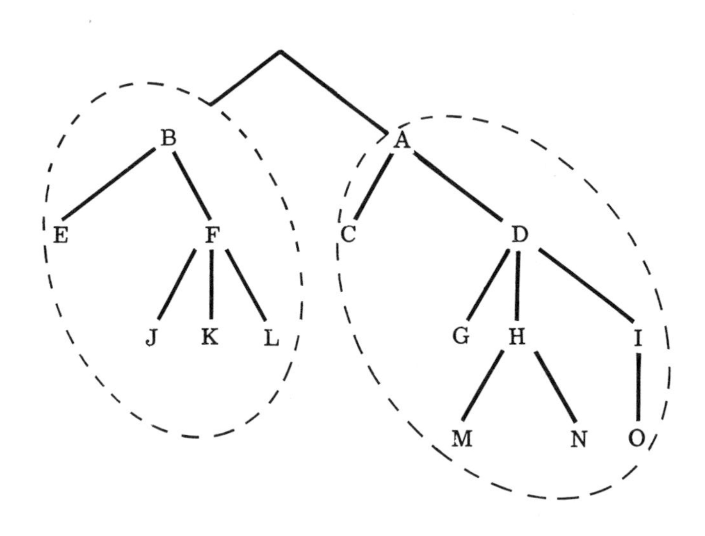


(b) The result of first cut between the root A and his eldest son B

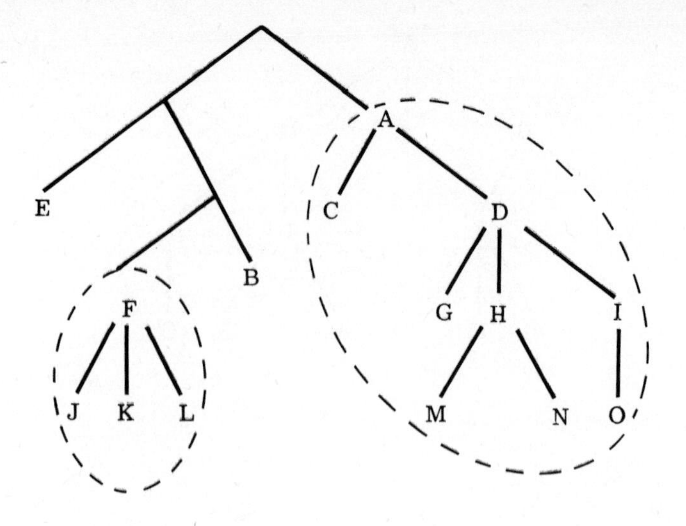


(c) The binary tree (still incomplete) after several further cuts

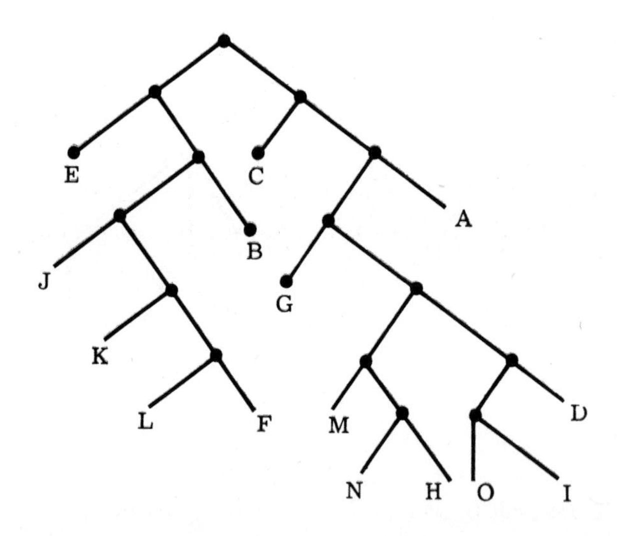


(d) The final binary tree

Fig. 6.9. The transformation of a tree of 15 nodes into a binary tree of 29 nodes

it down into a binary tree and the final result is shown in fig. 6. 9(d). In this final binary tree we see how the nodes of the original tree A - O have become the terminal nodes of the binary tree.

## 6. 4 TREE REPRESENTATION

Just as there are very many ways of traversing trees there are many methods of representing them in a computer and the choice of which representation to use in a particular application can make a great difference in computer time not merely a few per cent but often factors of two. In this section a selection of useful tree representations are described and in most cases an attempt has been made to give several typical examples of their uses. The reader would be well advised to try some of the sample problems using more than one method of tree representation. In this way one can obtain an insight into the strengths and weaknesses of these representations.

### 6. 4. 1 A Two-Link Method of Representing Binary Trees

Any node of a strictly binary tree or a Knuth binary tree has at most two subtrees. This means that there is a natural way to represent these binary trees with two links, a left link (LLINK) to the left subtree and a right link (RLINK) to the right subtree. A node can have a third field called DATA for the information at that node. (In section 6. 4. 2 we will consider another representation of a binary tree where the information is only held at the terminal nodes. Thus in these cases a branch node has no DATA field. ) We also have a link variable T, called the pointer to the tree, which contains the address of the root of the tree. If $T = \Lambda$ then the tree is empty.

Consider the binary tree shown in fig. 6. 10. This binary tree can be represented in the two link method as shown in fig. 6. 11.

As an example of how to manipulate this binary tree representation, consider the problem of examining each node in the tree to see whether it has a certain property. Thus we wish to traverse the binary tree and we can do so in any of the orderings we considered in section 6. 2. 1. Let us take symmetric order traversal; the algorithm uses an auxiliary stack

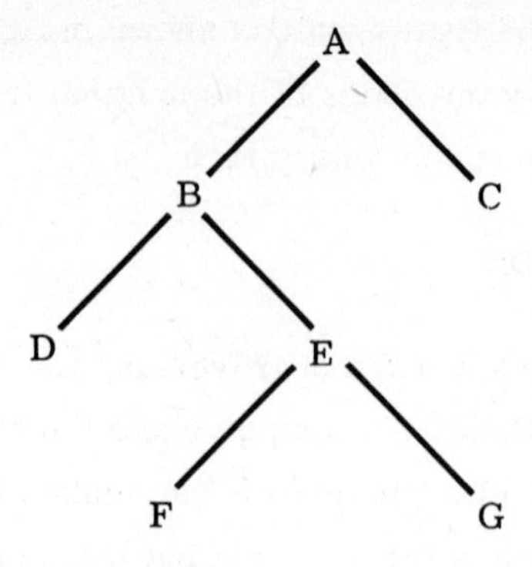


Fig. 6.10. A binary tree

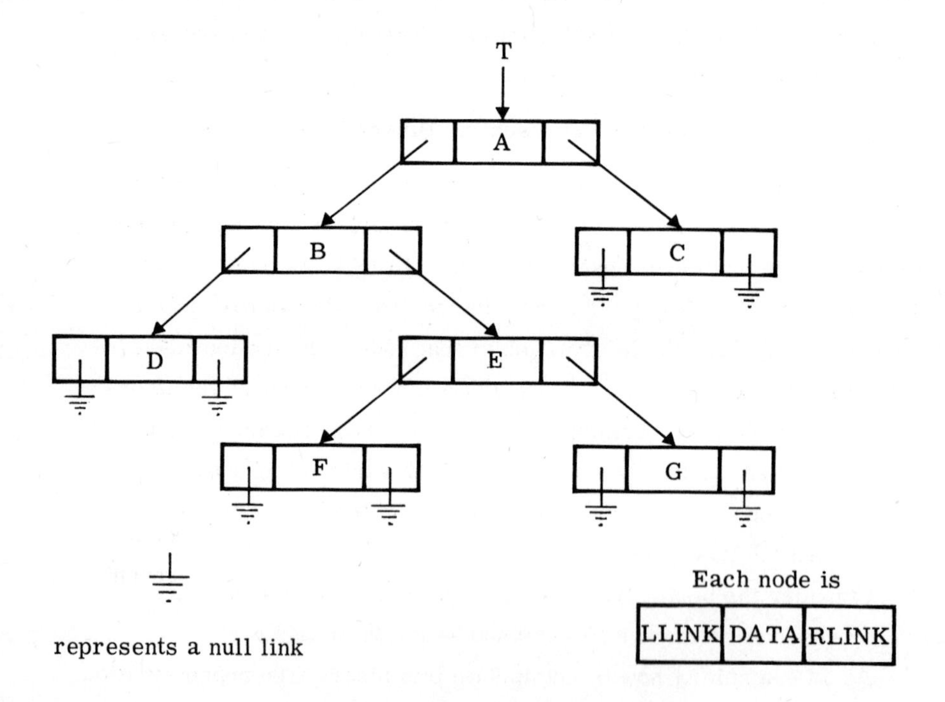


Fig. 6.11. The binary tree of fig. 6.10 in the two-link representation

S, which is assumed to be empty originally. The stack can be operated by the two procedures:

STACK (P, S); Stacks pointer P on the top of stack S.

UNSTACK (P, S, FAIL); Lets the pointer P become equal to the top element of the stack S unless stack S is empty in which case it jumps to the label FAIL.

The Algorithm for symmetric order traversal is

P:= T; **comment** T is the tree pointer;

LOOP: **if** P $\neq \Lambda$ **then begin**

STACK (P, S); P:= LLINK(P); **goto** LOOP **end**;

UNSTACK (P, S, END);

**comment** at this stage we examine the node P to see if it possesses the required property;

P:= RLINK (P); **goto** LOOP;

END:

For the binary tree in fig. 6.11 the above algorithm for symmetric order traversal works as shown in fig. 6.12.

| P | Stack | Action |
|---|---|---|
| A | A | Stack A |
| B | B, A | Stack B |
| D | D, B, A | Stack D |
| $\Lambda$ | B, A | Unstack D |
| $\Lambda$ | A | Unstack B |
| E | E, A | Stack E |
| F | F, E, A | Stack F |
| $\Lambda$ | E, A | Unstack F |
| $\Lambda$ | A | Unstack E |
| G | G, A | Stack G |
| $\Lambda$ | A | Unstack G |
| $\Lambda$ | - | Unstack A |
| C | C | Stack C |
| $\Lambda$ | - | Unstack C |
| $\Lambda$ | - | Unstack with an empty stack so algorithm terminates |

Fig. 6.12

Recursive methods are very convenient for traversing trees when the traversal is recursively defined. In the above example of symmetric order traversal of a binary tree with two links using ALGOL W and assuming a node of the binary tree is defined as

RECORD TREENODE(REAL DATA; REFERENCE (TREENODE)LLINK,RLINK);

the procedure would be

```
PROCEDURE SYMM_ORDER_TRAVERSAL (REFERENCE(TREENODE)
VALUE P);
IF P  ≠ NULL  THEN
 BEGIN SYMM_ORDER_TRAVERSAL (LLINK(P));
        VISIT(P);
        SYMM_ORDER_TRAVERSAL (RLINK(P));
 END;
```

In this procedure VISIT(P) is a procedure or piece of code that examines the node pointed to by reference P to see if it possesses the required property.

### 6.4.2 Other Methods of Representing Binary Trees

The two-link method given in the previous section is the most natural one for representing binary trees. However there are many other methods some of which will be discussed briefly in this section.

One method which has been used occasionally is to add a further link to make a three-link representation of a binary tree. This extra link points to the father of the node. Readers are advised to attempt example [6.13] at the end of this chapter which is a repeat of the problem of examining each node in the binary tree by symmetric order traversal using this new representation. The extra link means we do not need an auxiliary stack since we can use the link to the father to go back up the tree. We do however need to take precautions against visiting each node more than once. One method of doing this is to introduce a one-bit TAG, which typically could be the sign bit of one of the links, this TAG can be made negative when we visit the node for the first time and thus if we try to visit the node again a test on the TAG prevents us from doing it. Thus the extra link does not make it any easier to do this traverse, which illustrates that extra linkage does not always lead to a gain in efficiency. The work in manipulating extra links may outweigh the gain in mobility about the tree from those links.

A representation that is useful if array storage only is available is to consider the binary tree as a complete binary tree. That is all nodes are present at every level except possibly at the lowest level. This may mean that dummy nodes have to be added.

For example the binary tree in fig. 6.3(b) could be represented by a

**string array** TREE (1 :: 15)

and this array would contain the following values

| k | 1 | 2 | 3 | 4 | 5 | 6 | 7 | 8 | 9 | 10 | 11 | 12 | 13 | 14 | 15 |
|---|---|---|---|---|---|---|---|---|---|---|---|---|---|---|---|
| TREE(k) | A | B | C | D |  | E | F |  |  |  |  |  | G |  | H |

a blank string representing a dummy node.

A binary tree with n levels requires an array of size $2^n - 1$, which is the number of nodes in the complete binary tree. The left and right sons of the $k^{th}$ node are nodes $2k$ and $2k + 1$ respectively. For the example above the $6^{th}$ node E has nodes 12 and 13 as his sons, i.e. no left son and his right son is G.

Another interesting representation of binary trees occurs when there is information only at the terminal nodes. In the two-link representation of a binary tree given in section 6.4.1 there were three fields at each node, namely a LLINK, RLINK and DATA. This last field could either be the 'data' itself or an address leading to that data. In some applications the information is held just at the terminal nodes and so at a branch node it is only necessary to have two fields for the left and right links; at the terminal nodes no links are necessary so that the whole node can be used for the information. A further one-bit tag is necessary at each node to indicate whether it is branch or terminal. Thus the information at a node is either two links or data and a tag. In section 6.3.2 it was shown how a tree could be transformed into a strictly binary tree with the nodes of the original tree as the terminal nodes of the strictly binary tree. The resulting strictly binary tree is therefore ideally suited to the representation described above.

If such a tree is used as a dictionary, where the terminal information is in some lexicographical order it may well be worthwhile to have additional information in the form of a partial key word at each node. When searching this will help considerably to prove that a given entry does not exist and also to locate an entry in the dictionary without searching more than one terminal node. This of course is returning to the representation used in section 6.4.1 with a DATA field at all nodes. Such a method of searching corresponds of course to the 'binary chop' approach; first the half of the dictionary containing a given word is determined from a knowledge of the word at the middle point, then the correct quarter from the word at the appropriate quartile and so on.

### 6.4.3 A Simple Tree Representation

In this representation the nodes of the tree are supposed labelled 1 to n and for each node we record the number of the node immediately preceding it on the unique path from the root to that node. (This makes use of the theorem that there is a unique path between any two nodes in a tree.) For the node k this gives an integer which we will call 'above[k]' (sometimes called the 'label' of k). In Algol notation where the **integer** n gives the number of nodes, we would use the

**integer array** above[1:n];

to specify the tree. Clearly the root of the tree needs special attention: if the root is labelled by the integer r, then by convention we set above[r] = 0.

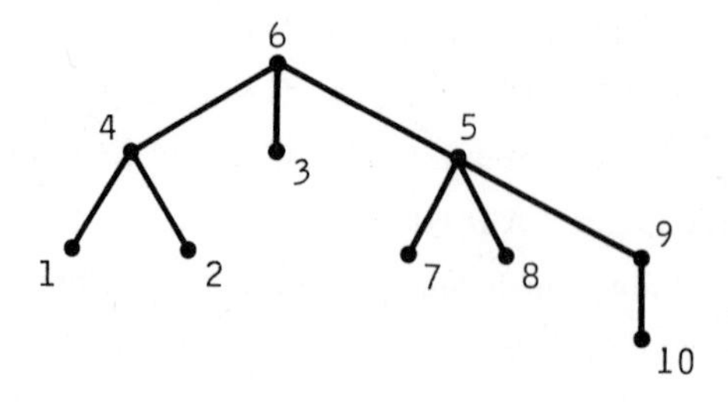


Fig. 6.13. A labelled tree with ten nodes

Given the tree shown in fig. 6.13, the array above would contain the following values

| k | 1 | 2 | 3 | 4 | 5 | 6 | 7 | 8 | 9 | 10 |
|---|---|---|---|---|---|---|---|---|---|---|
| above[k] | 4 | 4 | 6 | 6 | 6 | 0 | 5 | 5 | 5 | 9 |

The lines of the tree are given by

$$\{(k, \text{above}(k)) : k = 1, 2, \ldots r-1, r+1, \ldots n\}.$$

We have to take care to omit the line from the root, r, to its conventional father.

An example of a convenient use of this representation is finding a path from one node in the tree to another; the representation would be quite inconvenient if we had a problem which involved going down the tree towards the terminal nodes.

Three algorithms for finding the path in a tree from one node (p) to another (q) are given below.

<u>Algorithm A.</u> Ascend the tree from each node to the root keeping a record of the two paths taken. Go through both paths together starting at the root until the lowest common point is found.

<u>Algorithm B.</u> Ascend the tree from one node to the root keeping a record of the path (P) taken. Ascend the tree from the other node testing each new node passed through to see if it is on the path P. When such a node is reached the lowest common point has been found.

<u>Algorithm C.</u> This algorithm assumes the level number for each node is known. Trace back from node p or q with the highest level number until the two level numbers are the same. Now trace both branches towards the root together, level by level, until a common point is reached.

To find the path from, for example, node 7 to node 10 using Algorithm A we ascend the tree from node 7 to the root through the sequence 7, 5, 6, 0, where the next element in the sequence is supplied by the value in the array above. Similarly we can ascend from node 10 through the sequence 10, 9, 5, 6, 0, again stopping on the special integer 0 which is the father of the root. Going through both sequences together from the root we find 5 is the lowest common point and so the path from 7 to 10 is 7, 5, 9, 10.

Using the representation of this section Algorithm A is given below as a block of Algol program. We assume the tree has n nodes, and we wish to trace the path from node p to node q, and that a printing procedure called 'print line (r, s)' exists which prints r - s.

```
begin integer array list p, list q [1:n]; integer sp, sq, j;
        procedure trace (list, s, t); integer array list;
                integer t, s;
        comment this procedure traces the path from node t to the
        root, the nodes on this path are stored in list and the number of
        nodes on the path will be s;
        begin integer k; s:= 0, k:= t;
        while k ≠ 0 do
                begin s:= s+1; list[s]:= k;
                        k:= above [k]
                end
        end trace;
           trace (list p, sp, p); trace (list q, sq, q);
           comment we have now traced back from both the nodes p and
           q to the root and we must now go down from the root to find
           the lowest common point;
           L: if listp[sp] = listq[sq] then
                begin sp:= sp-1; sq:= sq - 1; if sp × sq ≠ 0
                then goto L end;
           comment finally we print out the path from p to q;
                for j:= 1 step 1 until sp do
                        print line (listp[j], listp[j+1]);
                for j:= sq step - 1 until 1 do
                        print line (listq[j+1], listq[j]);
end
```

The efficiency of the three different algorithms suggested for this problem would depend on the type of tree and the length of the path. For example Algorithms B and C would be more efficient than Algorithm A if the tree had a large number of nodes and the paths to be found were likely to be short.

The same representation can also be used for forests of trees since there is nothing that requires the label numbers of the nodes to belong to the same tree. One disadvantage of this representation is that ordering is not intrinsically included. For example in fig. 6.13 if nodes 1 and 2 were interchanged the above representation would be unaltered.

Another application of this representation is given in section 6.5.2 where a program for finding the minimum weighted path length of a binary tree using the Huffman algorithm is given.

### 6.4.4 Representing a Tree by a Terminating Binary Sequence

A very interesting representation of a tree has been introduced by Scoins in the form of a terminating binary sequence (tbs). We need to introduce another method of traversal of a binary tree, called reverse endorder traversal which can be defined recursively as follows:

Do nothing if the binary tree is empty,
Visit the root,
Traverse the right subtree in reverse endorder,
Traverse the left subtree in reverse endorder.

For a general tree we can construct the tbs representation as follows:

(i) Transform the tree into a strictly binary tree by the construction given in section 6.3.2.

(ii) Traverse the resulting strictly binary tree in reverse endorder and when we visit a node put a 1 in the binary sequence if the node is a branch node and a 0 if the node is a terminal node.

For example consider the tree given in fig. 6.9(a): this is transformed into the strictly binary tree shown in fig. 6.9(d). This resulting strictly binary tree has been redrawn below with the branch nodes labelled so that we can follow through the reverse endorder traversal. The reverse endorder traversal of the binary tree in fig. 6.14 visits the nodes in the order 1 3 5 **A** 7 9 12 **D** 14 **I O** 11 13 **H N M G C** 2 4 **B** 6 8 10 **F L K J E**.

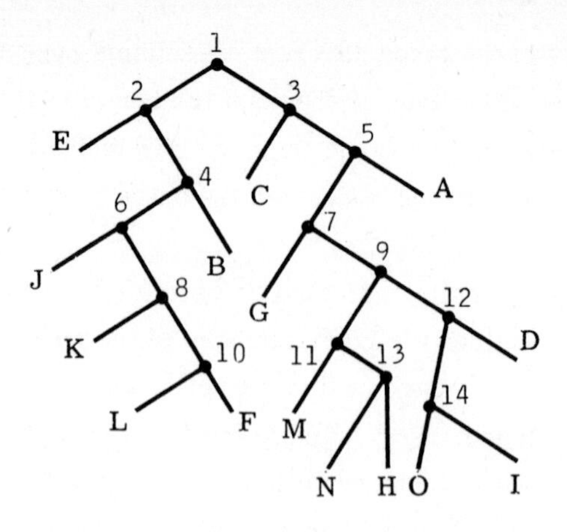


Fig. 6.14. The binary tree given in fig. 6.9(d) with the branch nodes labelled

If we now write the tbs by putting 1 for a branch node and 0 for a terminal node we get

$$111\ 0\ 111\ 0\ 1\ 00110\ 0\ 0\ 0\ 0\ 110111\ 00000\ . \qquad (6.3)$$

This then is the terminating binary sequence for the tree shown in fig. 6.9(a). It is somewhat more difficult to show that such a construction is unique but it is not too difficult to see that the construction can be reversed and from an allowable tbs we can obtain a tree.

There are some interesting properties of terminating binary sequences

(1) A tree with $n$ nodes has a tbs with $n$ 0s and $n - 1$ 1s.

(2) A group of $k$ 1s corresponds to a node with $k$ sons in the original tree.

(3) The zero that terminates the sequence of $k$ 1s corresponds to the node which has the $k$ sons.

For example, in the tbs given above the first three 1s indicate that the root has three sons. The next 0 corresponds to the root and the next three 1s indicate that node D has three sons etc.

The tbs given is unique for the tree given in fig. 6.9(a) under the method of construction described, but it is possible to obtain a different tbs for this tree by having a new construction method (e.g. traversing the nodes of the strictly binary tree in preorder or varying the methods of constructing the strictly binary tree from the original tree).

Although it is thought that the terminating binary sequence is a minimal representation of a tree with no information lost it is not necessarily convenient for all manipulation purposes. One possible use is to find in a given tree all the occurrences of a subtree with a certain structure.

Suppose we are looking for occurrences of the subtree structure shown in fig. 6.15. The tbs for the subtree structure in fig. 6.15 is

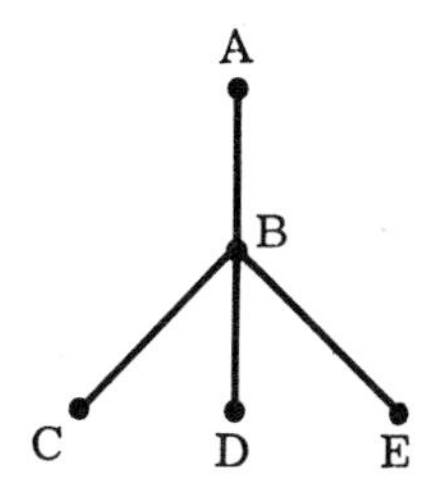


(a) Given subtree structure

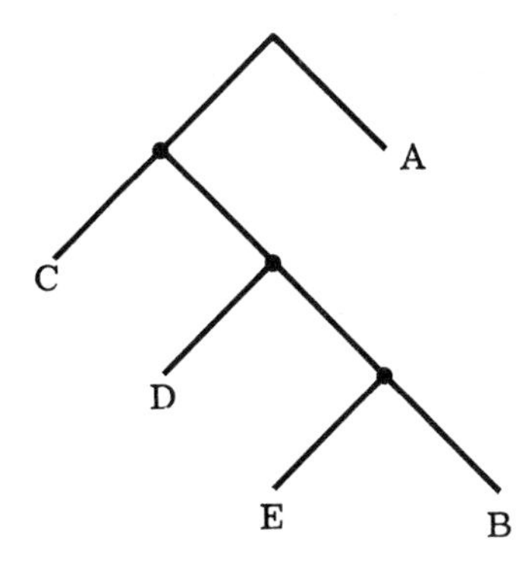


(b) Transformation of the above subtree into a binary tree

Fig. 6.15.

therefore

$$101110000 \, . \tag{6.4}$$

We can now find all occurrences of this subtree structure in the tree shown in fig. 6.9(a) by moving the binary pattern (6.4) through that given in (6.3) to see if it matches. In fact we do find one place where an exact match occurs which is the subtree

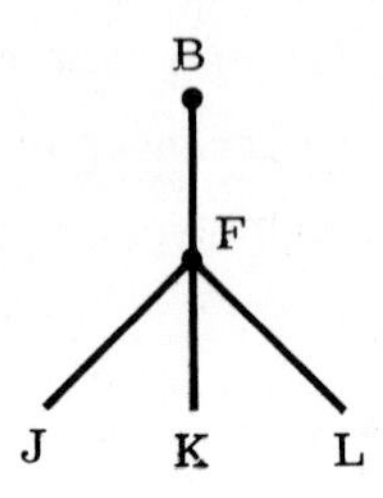


of the original tree. This of course is a very simple substructure matching and we might wish to consider

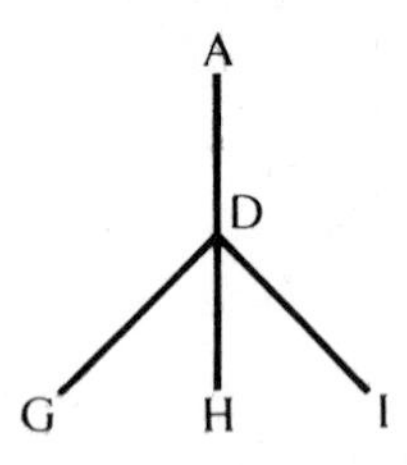


as another occurrence of the subtree. In this case we can still use the two tbs' (6.3) and (6.4) but the actual comparison algorithm is different.

### 6.4.5 A Representation of Trees using two Links and Tags

The two-link representation is an effective way of working with binary trees as we saw in section 6.4.1; however when we come to represent general trees we need a one-bit tag in addition to the two links. In practice it is often convenient to use the sign bit as the tag. The links and tag are therefore

FSON - an explicit link to first (or eldest) son.

BORF - a link to the right brother or if there is no right brother to the father.

STAG - a one bit tag which is + if the node is the youngest son and - if not.

Consider the tree with ten nodes shown in fig. 6.13: this can be drawn as in fig. 6.16. The table of links and tags for the tree shown in

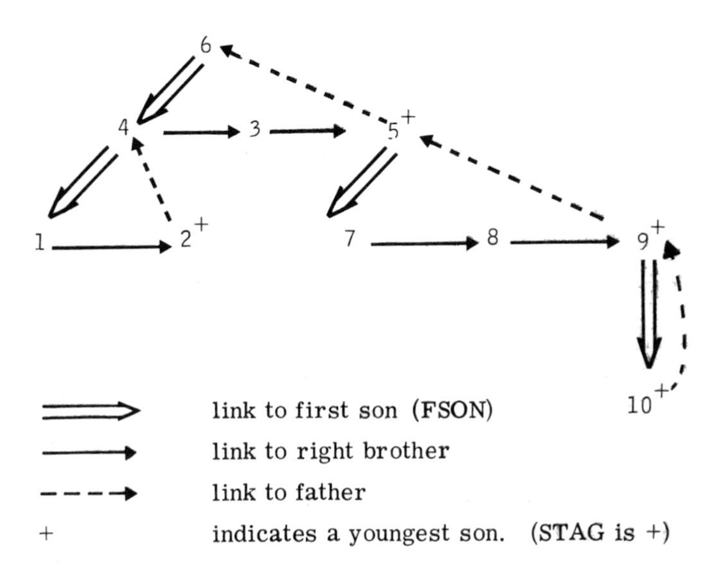


Fig. 6.16. A ten-node tree represented by two links and a tag

fig. 6.16 would be that of fig. 6.17.

| Node Label | FSON Link | Tag | BORF Link |
|---|---|---|---|
| 1 | - | - | 2 |
| 2 | - | + | 4 |
| 3 | - | - | 5 |
| 4 | 1 | - | 3 |
| 5 | 7 | + | 6 |
| 6 | 4 | + | 6* |
| 7 | - | - | 8 |
| 8 | - | - | 9 |
| 9 | 10 | + | 5 |
| 10 | - | + | 9 |

* We have conventionally let the BORF link of the root be the root itself.

Fig. 6.17

**Example.** Suppose we wish to search a tree represented in this way to find the father of the node whose data field is some unknown X.

We will assume each node n has a data field DATA(n) and if a node has no sons (i.e. it is a leaf) then FSON(n) = 0. The address of the root of the tree is in ROOT.

```
        n:= ROOT;
LOOP  : if DATA(n) ≠ X then
           begin comment traverse the tree in preorder;
                if FSON(n) ≠ 0 then n:= FSON(n) else
                begin comment there is no son.
                      As long as there is also no right brother go back to
                      the father. If the root is reached on the way then
                      the whole tree has been searched without finding X;
```

```
            for  n:= BORF(n)  while  n > 0  do
            if  n = ROOT  then goto  NOT FOUND;
            comment  now there is a right brother so we go to
            him;
            n:= -n
      end;
      goto  LOOP
end;
comment  we have found the node  n  whose data field is  X  and
we find its father as follows;
while  BORF(n) < 0  do  n:= -BORF(n);
father:= BORF(n);
```

Thus in this representation we find the father of a node by going down the right brothers until we reach the youngest son and then we find the father in the link BORF.

In this representation the FSON link will often be a null link, as we can see from fig. 6.17. In fact for a binary tree we can show that the number of terminal nodes is always one more than the number of branch nodes. So many null links wastes useful storage space which could be used in several ways, for example to make traversing the tree easier. We can achieve this by adding a one-bit tag (e.g. the sign digit) to the link. If the sign bit is positive then the link is a normal link to the first son, but if the sign bit is negative instead of the null link (since this is the case of no sons) we have a so-called thread link, which links a terminal node with some other node of the tree. Several possibilities are available and one that some authors think is the most satisfactory is to ascend the tree from node X until node Y is reached whose first son is different from the one just passed. Then the thread link goes from node X to node Y. For example consider the ten-node tree given in figs. 6.13 and 6.16. Nodes 1, 2, 3, 7, 8, 10 are terminal nodes and we can add a sign tag to them and turn their null links into thread links. The resulting thread links are shown in fig. 6.18.

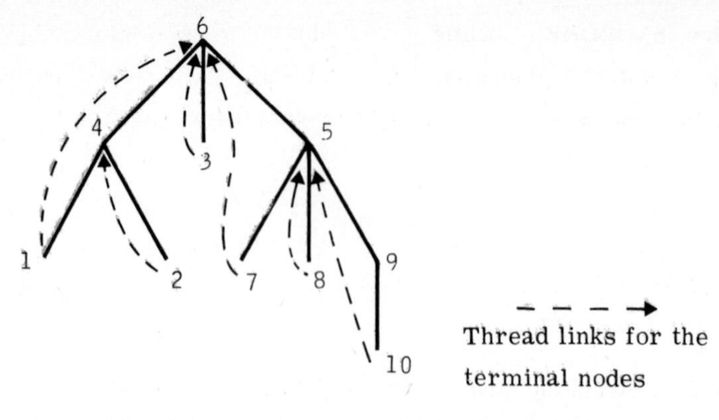


Fig. 6.18. A ten-node tree with its thread links

There is a special relationship between the leftmost terminal node (1 in the example above) and the root of the tree. This thread link does not therefore obey the general rule given above.

### 6.4.6 General Comments on Tree Representation

We have already seen in this chapter that there are many ways of representing trees in storage. The choice of the best representation, as always, depends on the operations needed in the problem to be solved. The ease of performing the operations will depend on the computer and programming language available as well as upon the representation selected. Most often a two-link representation seems to work well, particularly for binary trees, but representations with as many as four links (e.g. father, right brother, left brother, first son) have been suggested. A many-linked tree enables one to traverse easily in all directions and, in general, insertion and deletion are not difficult. However, apart from the extra storage required for the links, there may be considerable amount of rather unproductive link processing required. For example, consider the problem of finding in a given tree all occurrences of a subtree with a certain given structure. The most obvious method is to consider each node of the given tree in turn and check whether the links and tag bits of the subtree correspond in the correct order to the links and tag bits of the given tree. This can be quite complicated, particularly if there are many links and tag bits to check. The fewer the explicit links, the easier the search becomes. Therefore a representation like the terminating binary sequence given in section 6.4.4 is likely to be more convenient. As we found with linear lists, the most difficult problems are those where the representational requirements conflict. For example, we may wish to reduce the number of explicit links in order to do subtree matching in one part of the program and in another we may want a well-linked tree to ease inserting and deleting. The art of computer science is often that of finding good compromises!

## 6.5 PATH LENGTH

The path length of a tree is the sum of the lengths from the root to each node. The path length is therefore important when using trees in computer algorithms since it is very closely related to the execution time. Thus if there are n nodes in the tree

$$\text{Path length} = \sum_{i=1}^{n} l_i$$

where $l_i$ is the distance from the ith node to the root.

### 6.5.1 Path Length for a Binary Tree

We have seen that the binary tree is a very important special case and it is often convenient to define two path lengths for such a tree. These two path lengths are called the external path length and the internal path length and are defined as follows.

Given any strictly binary tree (a similar construction can be applied to a Knuth binary tree), let the nodes be represented by circles. At each terminal node add two special nodes (represented by squares). Thus given the strictly binary tree shown in fig. 6.19(a) we transform it into that given in fig. 6.19(b).

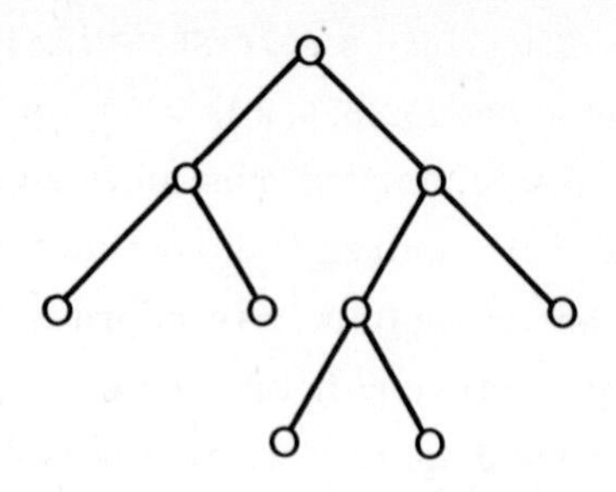

(a) The original strictly binary tree

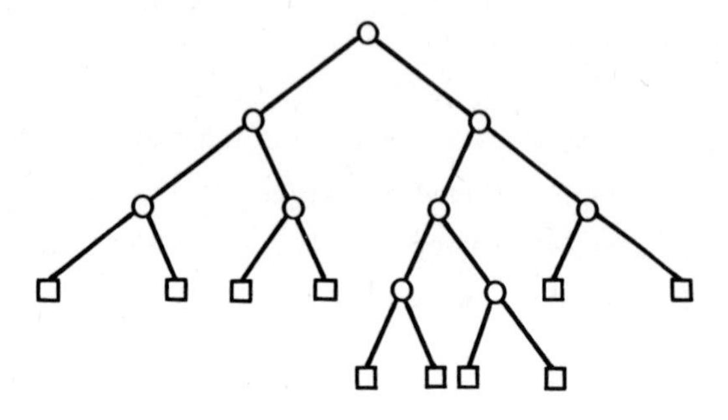

(b) The transformed binary tree

Fig. 6.19.

If there are n nodes in the original binary tree we can show there are $2n + 1$ nodes in the transformed binary tree and thus the number of square nodes, s is equal to $n + 1$.

The transformed binary tree is often theoretically more convenient and we will work with it.

The external path length, E, is defined as the sum of the lengths of the paths from the root to all the terminal (square) nodes. The internal path length, I, is defined as the sum of the lengths of the paths from the root to all the branch (round) nodes. For the example given in fig. 6.19,

$$E = 3 + 3 + 3 + 3 + 4 + 4 + 4 + 4 + 3 + 3 = 34,$$
$$I = 0 + 1 + 1 + 2 + 2 + 2 + 2 + 3 + 3 = 16\ .$$

In general we can show that the two quantities E and I are connected by the relation

$$E = I + 2n . \qquad (6.5)$$

The binary tree with minimum path length can be found without too much difficulty. Thus the binary tree with n nodes having minimum path length has a root, two nodes at level 2, four nodes at level 3 etc. At the lowest level there may not be as many as $2^i$ nodes. Such a tree is called a complete binary tree for n nodes. A complete binary tree for 9 nodes is illustrated in fig. 6.19. By summing the series

$$0 + 1 + 1 + 2 + 2 + 2 + 2 + 3 + 3 + 3 + 3 + 3 + 3 + 3 + 3 + 4 + 4 + \qquad (6.6)$$

we can find the internal path length of a complete binary tree and it will be essentially of the form nlogn (see section 7.2.2).

### 6.5.2 Weighted Path Length for a Binary Tree

We can extend the ideas in the previous section by giving each node a weight, $w_i$, and instead of minimizing $\sum_i l_i$ we minimize $\sum_i w_i l_i$.

This is called the weighted path length. Often it is convenient to have weights only at the terminal nodes and so we need only some of these nodes. For example consider the trees given in fig. 6.20.

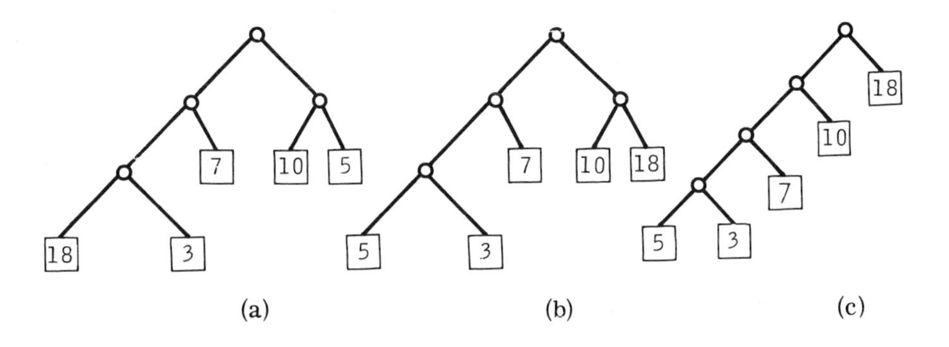


Fig. 6.20. Binary trees with five weights at the terminal nodes

The weighted path lengths for the trees given in figs. 6.20(a), (b) and (c) are respectively 107, 94, 91. When the nodes are weighted the complete binary tree as shown in figs. 6.20(a) and (b) does not necessarily give the minimum weighted path length. In fact in case (b) the smallest weights have been placed furthest from the root of the complete binary tree.

The method of finding the minimum weighted path length brings us back to the Huffman algorithm first introduced in section 2.6.1 in connection with finding optimal codes given the probabilities. In fact the algorithm is exactly the same, e.g. if we are given the weights 18, 10, 7, 5, 3, we combine the two smallest 5 + 3 and then recursively work on the problem with one less weight i.e. 18, 10, 8, 7.

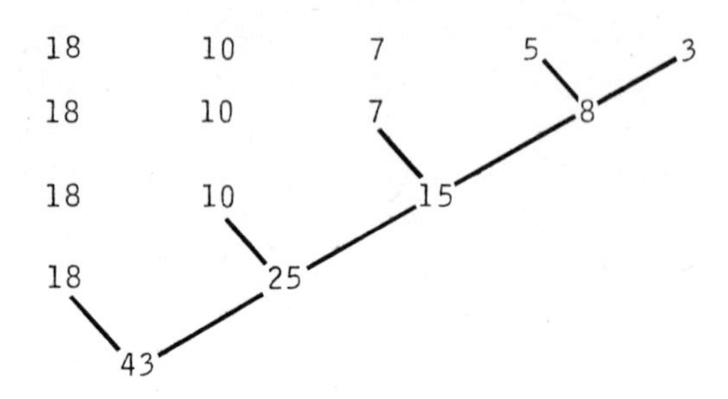


The resulting tree is in fact shown in fig. 6.20(c).

In order to illustrate the use of tree structures and tree representations we will write a program using the Huffman algorithm to find the tree for the minimum weighted path length. The tree representation used is that given in section 6.4.3 where the entry in the array 'above' refers to the father of the node.

```
procedure HUFFTREE (n, max, weight, above); value n, max;
integer n; real max; integer array above; real array weight;
comment the n weights are in the array weight and 'max' is a
value greater that than the sum of all the weights. 'above' is an
array [1:2 × n - 1] which will on exit hold the links of the tree,
such that above[i] is the father of node i;
begin real min1, min2; integer i, j1, j2, k;
        for i:= 1 step 1 until 2 × n - 1 do above[i]:= 0;
        for k:= n + 1 step 1 until 2 × n - 1 do
        begin min1:= min2:= max; j1:= j2:= 0;
```

```
          for i:= 1 step 1 until k - 1 do
          if above[i] = 0 then
          begin comment search the weights for the two smallest;
                if weight[i] < min1 then
                begin min2:= min1; min1:= weight[i];
                      j2:= j1; j1:= i
                end
                else if weight[i] < min2 then
                begin min2:= weight[i]; j2:= i
                end
          end i;
          comment combine two smallest weights in min1 and min2
          and put them in the next available space;
          weight[k]:= min1 + min2;
          above[j1]:= above[j2]:= k
    end
end HUFFTREE;
```

At the termination of this procedure the minimum weighted tree is in the array above.

## 6.6 BIBLIOGRAPHY

The mathematical literature on trees is very extensive but most of it is only of marginal interest to a computer scientist. Basic theory is covered in Knuth [1] and Riordan [15]. Extensive and more detailed treatment of the material of this chapter are to be found in Knuth vol. 1 [1] and in two papers published in Machine Intelligence by Scoins [2], [3]. Part of the treatment is derived from a series of lectures by T. E. Cheatham, the essence of which is contained somewhat obscurely in [4]. Some other books and articles on trees and their application to computer science are Iverson [5], Johnson [6], D'Imperio [7], Braden and Perlis [8], Forsythe et alia [9], Berztiss [10], Elson [11], Korfhage [12], Shave [13], and Stone and Siewiorek [14].

[1] D. E. Knuth: The Art of Computer Programming, vol. 1: Fundamental Algorithms, pp. 305-406. Addison-Wesley, 1968.

[2] H. I. Scoins: Linear Graphs and Trees in Machine Intelligence, vol. 1, pp. 3-15, edited by N. L. Collins and D. Michie. Oliver and Boyd, 1967.

[3] H. I. Scoins: Placing Trees in Lexicographic Order in Machine Intelligence, vol. 3, pp. 43-60, edited by D. Michie. Edinburgh University Press, 1968.

[4] T. E. Cheatham: Notes for a course on 'Syntax Oriented Translations, Core Languages etc.' given at International Seminar on Advanced Programming Systems (vol. 2 Advanced Course) at the Hebrew University of Jerusalem, July 1968. Published by ILTAM.

[5] K. E. Iverson: A Programming Language. John Wiley and Sons, 1962.

[6] L. R. Johnson: System Structure in Data, Programs and Computers'. Prentice Hall, Inc., 1970.

[7] M. E. D'Imperio: Data Structures and their representation in storage in Annual Review in Automatic Programming, vol. 5, pp. 1-75, edited by M. I. Halpern and C. J. Shaw. Pergamon Press, 1969.

[8] R. T. Braden and A. J. Perlis: An Introductory Course in Computer Programming, pp. 53-65. D-S-C Monograph No. 7, Carnegie Institute of Technology, 15th June 1965.

[9] A. I. Forsythe, T. E. Keenan, E. I. Organick, W. Stenberg: Computer Science: A First Course, pp. 372-400, 418-50. John Wiley and Sons 1969.

[10] A. T. Berztiss: Data Structures - Theory and Practice, 2nd edition, Academic Press, 1975.

[11] Mark Elson: Data Structures, Science Research Associates, 1975.

[12] R. R. Korfhage: Discrete Computational Structures, Academic Press, 1974.

[13] M. J. R. Shave: Data Structures, McGraw-Hill, London, 1975.

[14] H. S. Stone and D. P. Siewiorek: Introduction to Computer Organisation and Data Structures: PDP11 Edition, McGraw-Hill, New York, 1975.

[15] J. Riordan: An Introduction to Combinatorial Analysis, pp. 107-62. John Wiley and Sons, 1958.

## EXAMPLES 6

[6. 1] Using either of the two definitions given in the text prove that a tree with n nodes has n - 1 lines.

[6. 2] Consider the tree shown in fig. 6. 21. Answer the following questions about this tree:

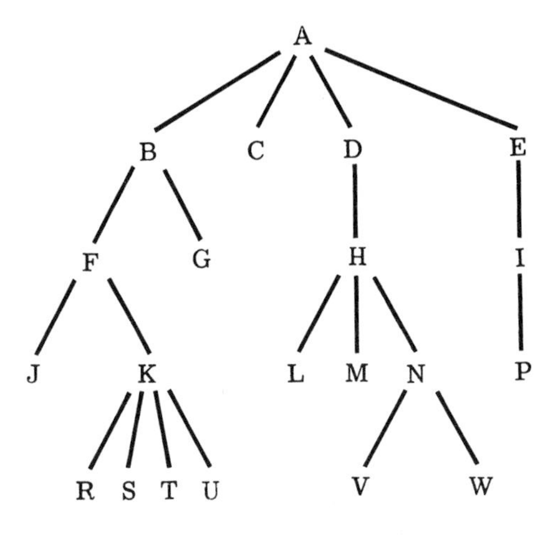


Fig. 6. 21. A rooted tree

(a) How many branch nodes are there?

(b) What is the relationship between U and J and between U and G?

(c) What is the degree and level of the nodes A, N and R?

[6. 3] Represent the following three arithmetic expressions as trees

$a^5 + 5a^3 - 2a + 7,$

$a/b/c + d \times e \times \sin(f + g) + \log t,$

$s^{a+b^n}.$

[6. 4] Draw a decision tree for the 'eight-coin' problem. The problem is given eight identical looking coins and an ordinary equal arm balance, identify in three weighings the one coin that is either heavier or lighter than the others?

[6. 5] Traverse the tree in fig. 6. 21 in the following ways

(a) Bottom up. Level-by-level.

(b) Suffix walk.

(c) Family order.

Can it be traversed in reverse endorder?

[6. 6] Transform the tree in fig. 6. 21 into

(a) A Knuth binary tree,

(b) A strictly binary tree.

[6. 7] Traverse in preorder both the tree in fig. 6. 21 and its transformation into a Knuth binary tree.

[6. 8] Write detailed Algol programs to find the path between nodes p and q in a tree using the simple 'above' representation and Algorithms B and C of section 6. 4. 3.

[6. 9] Give the terminating binary sequence (tbs) representation of the tree in fig. 6. 21.

[6. 10] If the tree in fig. 6. 21 were represented in the form given in section 6. 4. 5 give a table showing the FSON and BORF links and the sign tag.

[6. 11] Write an algorithm to traverse in family order a tree represented as in section 6. 4. 5.

[6. 12] Evaluate the equation (6. 6) to find the internal path length of a complete binary tree with n nodes.

[6. 13] Write an algorithm to traverse in symmetric order a binary tree represented by the three-link method of section 6. 4. 2.

[6.14] Given the following 'threaded list' data structure, exhibit the same data as (i) a tree, and (ii) as nested parenthesized sequences, such as ((a, b), c):

| | |
|---|---|
| 1 | 1, 2, 0 |
| 2 | 0, p, 3 |
| 3 | 1, 4, 6 |
| 4 | 0, q, 5 |
| 5 | 2, r, 3 |
| 6 | 3, 7, 1 |
| 7 | 0, s, 8 |
| 8 | 3, 9, 6 |
| 9 | 0, t, 10 |
| 10 | 2, u, 8 |

Suppose a new kind of entry (type 4) were needed to point back to the head of the list. Assume it has the form, 4, 1, $\gamma$ where $\gamma$ is its successor. It would be a good idea to insert such an item fairly often, and we stipulate that it should follow (logically, not physically) the last item in each sublist. For example, it would be needed at the place indicated in the following example:

((a, b ), c ).
↑ ↑

(a) Use the 10-entry list given above to illustrate the change. Put new type 4 entries in as entries at 11, 12 and so on, making as few changes in earlier entries as possible, and circling all changes you do make.

(b) State briefly an algorithm which would make the changes you made in (a) in one pass. Assume the original length (in this case, 10) is known as the value of a variable (L).

(Carnegie Institute of Technology, 1967)

[6.15] Give details of a simple scheme for the symbolic differentiation of algebraic expressions. Show graphically how the data structure develops during the differentiation of the expression $y = x + 8$.

(Glasgow 1969)

[6.16] Draw binary trees to represent the structure of the expressions

(i) $a * b - c/d$

(ii) $a + b + c + d$

(iii) $(a - b) * (c + d) \uparrow 2$

(iv) $A[i, j, k] - \sin(2*\pi*x)$,

and state precisely what assumptions you have made. Using the rule 'root, then branches anticlockwise' deduce the Polish prefix form of each expression.

Draw a flow diagram to indicate how the Polish prefix representation of a tree may be used to recover the original form of the expression.

For which expressions would a non-binary tree be a more convenient representation? Indicate how the Polish prefix form can be generalized to non-binary trees.

(Essex 1971)

[6.17] Show how a tree data structure can be represented by means of chained files. Discuss the facilities desirable in a data management system designed to manipulate such files.

(Manchester 1971)

[6.18] Distinguish between a directed graph and an undirected graph. How may a directed graph $G_d$ with vertices $V_1, V_2, \ldots V_n$ be represented as an ALGOL data structure, when there is at most one edge between any pair of vertices? Can you reduce the computer storage needed to hold this structure in the case when the graph $(G_u)$ is undirected but otherwise similar? Explain your reasoning.

Assuming each individual edge to be of unit length, outline how you would program a computer to find the length of the shortest path between any two vertices of $G_u$.

(Manchester 1971)

[6.19] A dictionary of identifiers can be stored as a tree, so that all identifiers starting with a given sequence of letters are to be found at or below a corresponding node of the tree. Thus, given that the following identifiers have been declared:

ABA

ABC

CAD

ABCD

he tree stored would appear as in fig. 6.22.

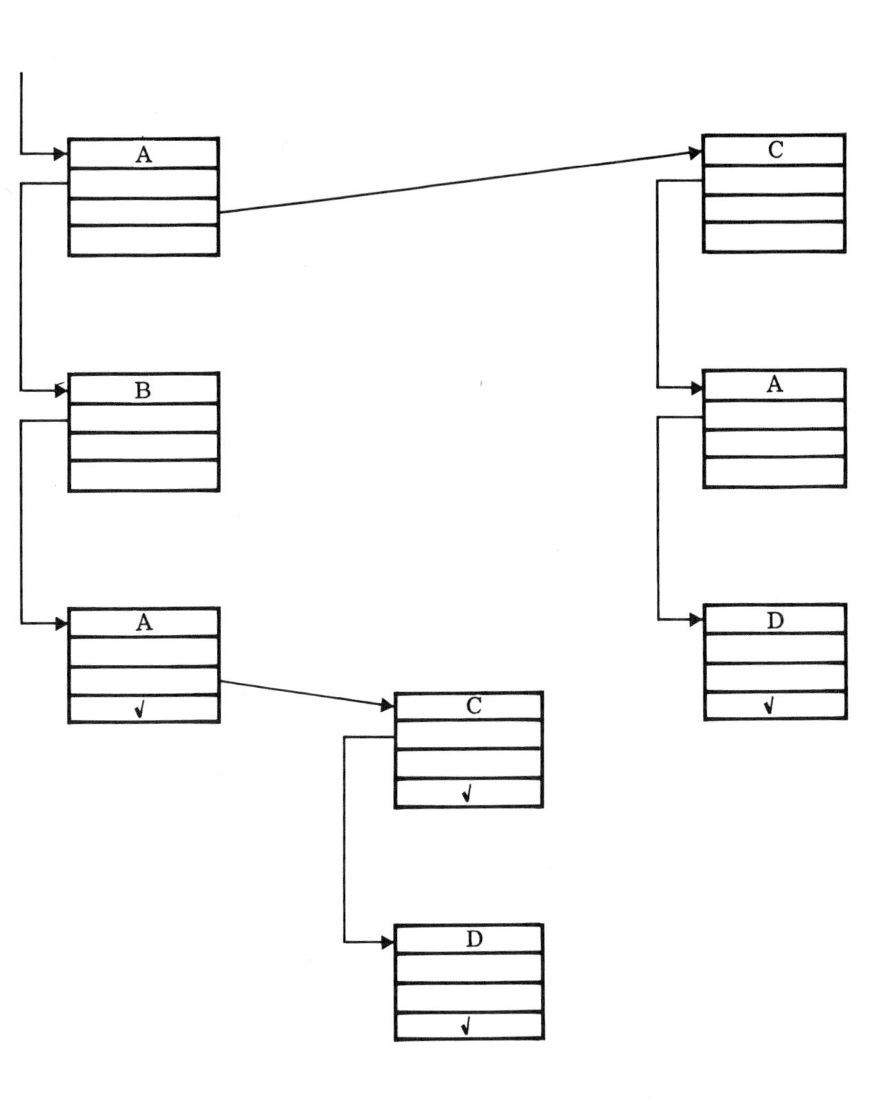


Fig. 6.22.

Each node contains two pointers, one to its daughter node and one to its sister node.

The ticks in the last field of each node are to indicate that a complete identifier ends there.

Write a clear and efficient procedure in ALGOL 68 (or PL/1, if you prefer) to add a new string (or CHARACTER (*)) to such a tree (unless it was there already). Incorporate this procedure into a program which will read and add sequences of identifiers.

Your program is expected to illustrate your knowledge of the essential features of the chosen language. Small syntactic errors will not therefore be penalized.

[In ALGOL 68, a string s may be read by 'read(s)', its length may be found by 'upb s', its ith char by 's[i]' and a substring from characters i to j inclusive by 's[i:j]'.]

[In PL/1, a CHARACTER VARIABLE S may be read by GET LIST (S), its length may be found by LENGTH(S), its Ith CHARACTER by SUBSTR (S, I, 1) and a substring of K characters starting at the Ith by SUBSTR(S, I, K). The address of a variable V may be obtained by ADDR (V).]

(Manchester 1971)

[6.20] Define the following terms:

(a) tree, (b) binary tree, (c) forest.

How may binary trees be represented internally in a computer? Show that there is a 1-1 correspondence between forests and binary trees and hence give a method for storing forests internally.

Write an algorithm which traverses a forest in preorder.

(St. Andrews 1971)

[6.21] A forest of two rooted trees is represented in a computer as in fig. 6.23.

Link 1 is to the first son (Link = 0 if no sons).

Link 2 is to the right brother if tag = 0.

Link 2 is to the father if the tag = 1 (Link = 0 if no father i.e. root).

Tag is 1 if it is the youngest son and 0 otherwise.

| Store location | Data | Link 1 | Link 2 | Tag |
|---|---|---|---|---|
| 100 | $A_1$ | 101 | 0 | 1 |
| 101 | $A_2$ | 0 | 102 | 0 |
| 102 | $A_3$ | 104 | 103 | 0 |
| 103 | $A_4$ | 106 | 100 | 1 |
| 104 | $A_5$ | 0 | 105 | 0 |
| 105 | $A_6$ | 0 | 102 | 1 |
| 106 | $A_7$ | 0 | 103 | 1 |
| 107 | $B_1$ | 108 | 0 | 1 |
| 108 | $B_2$ | 110 | 109 | 0 |
| 109 | $B_3$ | 112 | 107 | 1 |
| 110 | $B_4$ | 0 | 111 | 0 |
| 111 | $B_5$ | 0 | 108 | 1 |
| 112 | $B_6$ | 113 | 109 | 1 |
| 113 | $B_7$ | 0 | 114 | 0 |
| 114 | $B_8$ | 0 | 115 | 0 |
| 115 | $B_9$ | 0 | 116 | 0 |
| 116 | $B_{10}$ | 0 | 112 | 1 |

Fig. 6.23

Draw the trees $\{A_i\}$ and $\{B_i\}$ of this forest and also show how they can be represented as nested parentheses.

Describe one general way in which a forest can be transformed into a binary tree and draw the binary tree which corresponds to the forest $\{A_i\}$, $\{B_i\}$.

Define preorder and endorder traversal of a binary tree and show the order in which the nodes of the above binary tree would be visited in each case.

(Newcastle 1970)

[6.22] Describe how a general tree can be represented by a terminating binary sequence (t.b.s.).

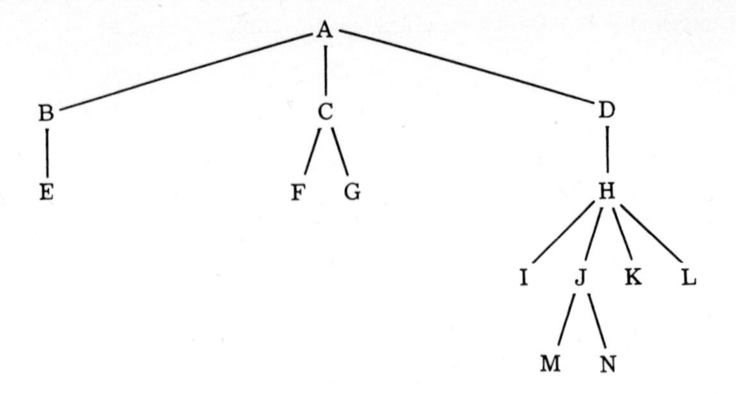


Show the process by which this tree is represented by a t. b. s. and determine the appropriate t. b. s. Prove the following

(a) A tree with $n$ nodes has a t. b. s. with $n$ zeros and $n - 1$ ones.

(b) A group of $k$ ones in the t. b. s. corresponds to a node with $k$ sons in the original tree. The zero that terminates the sequence of $k$ ones corresponds to the node which has the $k$ sons.

(Waterloo, 1974)

[6.23] <u>Programming Project</u>

Write a program to test the occurrences of a given tree substructure in a given tree. The trees may be represented as terminating binary sequences or in any other way you wish.

In particular consider what you mean by a tree substructure belonging to a tree and allow a substructure of the form

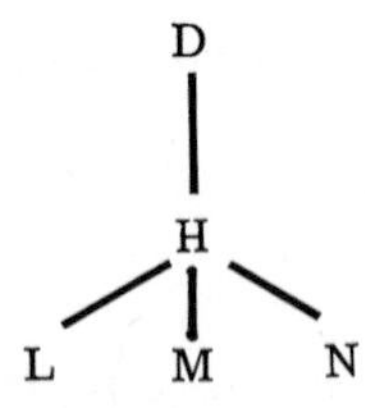


to occur in the tree of fig. 6.21.

[6.24] Programming Project: Expression Evaluation

Write a procedure which will read in an arithmetic expression and construct the corresponding 'operator tree', e.g.

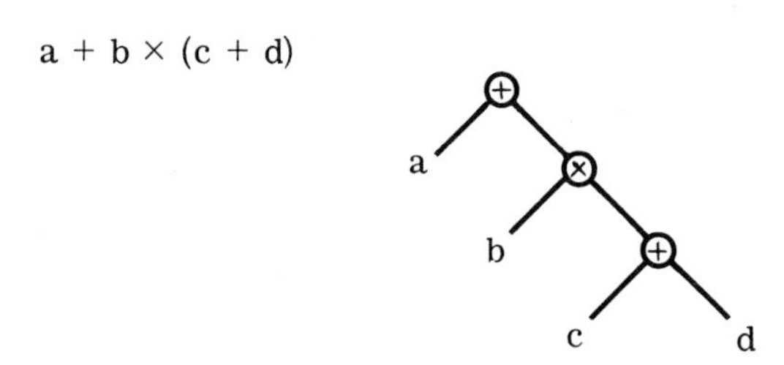


where the non-terminal nodes are labelled by the operators which operate on the two subtrees. Note that you must take into account the priority of the operators and the parentheses in the expression. Check your procedure by incorporating it in a program to print out the tree in pre-order (prefix notation) e.g. for the above $+ a \times b + cd$.

Write a program which uses this procedure to evaluate arithmetic expressions by constructing the tree and associated variable table for the expression read. Next, the program should prompt for values of the variables which is probably best done by calling the procedure again thus allowing, in general, that the value given to a variable may be an expression, in particular a number.

e.g.

```
            READY!
    X + 2 × Y
            WHAT IS  X?
    2
            WHAT IS  Y?
    A + B
            WHAT IS  A?
    1
            WHAT IS  B?
    X
            X + 2 × Y = 8
            READY!
```

Possible extension (i) By allowing variables to remain unknown, the result of evaluation could in general be an expression; thus for the example above, when the computer prompts for the value of A, if we had been non-committal, typing '?' in reply, the result could come

$$X + 2 \times Y = 2 + 2 \times (A + 2) .$$

(ii) Incorporate a simple command language. For example, instead of the program just prompting for values, you can enter a command, e. g.

$$SX = A + B + 1$$

meaning set X to value of A + B + 1.

(Reference: Knuth, vol. 1, pp. 335-347.)

[6.25] Programming Project: Propositional Calculus 1 (Tautologies)

For boolean variables p and q we denote by $(p \supset q)$ the proposition 'p implies q' which has the value false (F) only if p is true (T) and q is false:

| | $\supset$ | q: T | F |
|---|---|---|---|
| p | T | T | F |
| | F | T | T |

Using ~p to denote 'not' p where ~F = T and ~T = F we can define a set of formulae by the following syntax:

$$\langle exp\rangle ::= \langle var\rangle \,|\, (\langle exp\rangle \supset \langle exp\rangle) \,|\, {\sim}\langle exp\rangle$$
$$\langle var\rangle ::= a\,|\,b\,|\ldots|\,x\,|\,y\,|\,z\ .$$

Example: $(p \supset (p \supset {\sim}q))$.

Now it is possible to determine the values of such a formula for all values of the variables it contains:

| p | q | ~q | $(p \supset \sim q)$ | $(p \supset (p \supset \sim q))$ |
|---|---|---|---|---|
| T | T | F | F | F |
| T | F | T | T | T |
| F | T | F | T | T |
| F | F | T | T | T |

If it happens that a formula is true for all values of its variables then it is called a tautology. E.g. $(p \supset p)$, $(F \supset p)$:

| p | q | $(p \supset q)$ | $(q \supset (p \supset q)$ |
|---|---|---|---|
| T | T | T | T |
| T | F | F | T |
| F | T | T | T |
| F | F | T | T |

Write a program which will test if a formula is a tautology.

Note. Investigate the tree of a formula

e.g. $((p \supset q) \supset (q \supset p))$

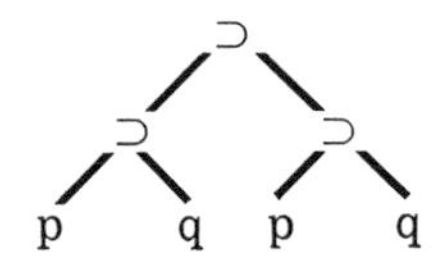


Extension. Consider how it might be possible to generate all tautologies in some order.

(Reference: Logic and Algorithms, Korfhage; chapter 3. Note that the notation there varies slightly from that used above.)

[6.26] Programming Project: Propositional Calculus 2 (Simplification)

Boolean variables p and q can be combined into propositions by various logical connectives. In particular we use $(p \wedge q)$, $(p \vee q)$, $(p \supset q)$, $(p \equiv q)$ to denote conjunction ('and'), disjunction ('or'), implication ('implies'), and equivalence, with the following meaning:

| p | q | $(p \wedge q)$ | $(p \vee q)$ | $(p \supset q)$ | $(p \equiv q)$ |
|---|---|---|---|---|---|
| T | T | T | T | T | T |
| T | F | F | T | F | F |
| F | T | F | T | T | F |
| F | F | F | F | T | T |

(for others see Korfhage).

Also we use $\sim p$ ('not' p), where $\sim T = F$, $\sim F = T$.

It is possible, however, to determine 'minimal' sets of connectives, in the sense that the other connectives can be determined by them.

e.g. $\{\wedge, \vee, \sim\}$

$(p \supset q)$ $\quad (\sim p \vee q)$

$(p \equiv q)$ $\quad (p \wedge q) \vee (\sim p \wedge \sim q)$

An interesting 'minimal' set of connectives is $\{\supset, \sim\}$.

By constructing the tree of a general formula (q. v.) it should be possible to translate this into an equivalent formula using a minimal set of connectives. Write a program to perform such a translation for the set $\{\supset, \sim\}$.

Extension. Consider the problem of performing the translation in the other direction where the goal is the least number of occurrences of the variables (or some other measure of conciseness of the expression).

(Reference: Logic and Algorithms, Korfhage, chapter 3.)

## [6.27] Programming Project: Draw a Tree

Write a program which will produce on a lineprinter a graphical representation of a tree. This representation should be as readable as possible. An example of such a representation might be as follows:

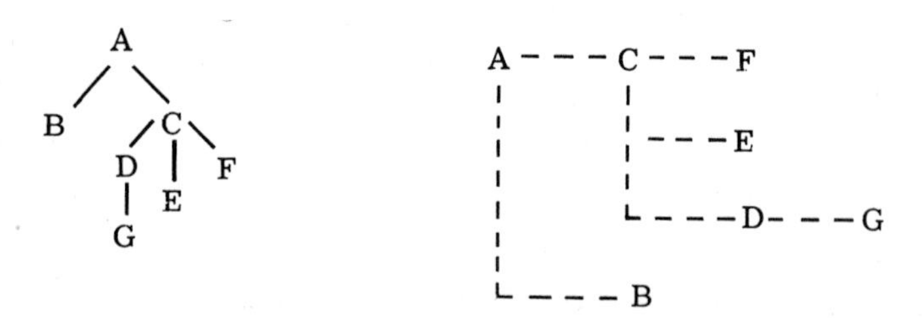


Alternatives are shown below:

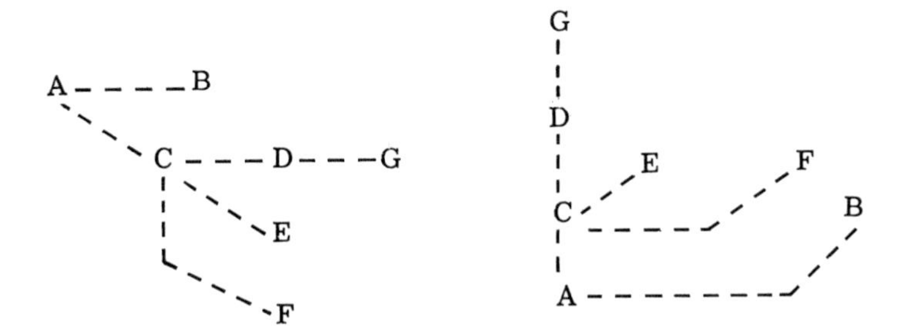


The representation should be chosen to simplify the programming initially.

# 7 · Searching

## 7.1 INTRODUCTION

The problem in searching is to locate a particular record in a given data set if such a record exists. The method of searching will depend upon the way the data is structured and represented, while at an earlier stage, the representation will have been chosen to take account of the kind and frequency of searching required. In this chapter each record will be considered to have two fields, the Key field and the Information field. These are sometimes known as the 'argument' and 'function value' respectively to give an analogy between the search operation and function evaluation. The key field is the one we search, and if it is convenient for manipulation, the key may be detached from the full record, so that the information field contains a pointer to the place where the full record may be found. (See also section 8. 1. 2 on Detached Keys. ) The record may also contain links if we are dealing with linked or semi-linked structures. In our searching algorithms, we will only explicitly consider the key field.

If the structures are dynamically varying, the searching operation must be considered in conjunction with the insert and delete operations. For example, we usually require algorithms which insert a new record in its correct position in the structure if the search operation fails to locate the required key.

We will classify searching algorithms into

(a) Scanning which involves comparisons between keys of the structure and the key being sought

(b) Key Transformation which involves finding the correct location (partially or totally) from some calculation based on the key itself. This technique is sometimes known as hash coding and scatter storage.

## 7.2 SCANNING

The scanning algorithms involve sequences of comparisons and it is useful to have a lower limit on the number of comparisons required. Each comparison between the sought for key and one of the $n$ keys in the structure yields one bit of information - the information that the key is either the correct one or not. Since there are $n$ keys, we know that to enumerate them in the binary scale would need approximately $\log_2 n$ bits, so that, on average, we expect from information theory that the number of comparisons in a scanning algorithm will be proportional to $\log_2 n$ (see section 2.5). Thus any general purpose scanning algorithm which uses about $C \log_2 n$ comparisons (where $C$ is a constant) can only be improved by reducing the value of the constant $C$. We can use this as a yardstick to measure the scanning algorithms.

Scanning algorithms can be divided into two classes called directed and controlled.

### 7.2.1 Directed Scanning

A directed scan is quite simply a search either from the beginning to the end or in the reverse order through a list of elements which may be linked or sequential.

The simplest case is searching an unordered list of elements in the array $a_1, a_2, \ldots a_n$ for some unknown element $x$. In this simple searching we compare each element in the list with the unknown $x$ as in the following Algol statement:

```
for i:= 1 step 1 until n do
if x = a[i] then goto FOUND;
```

if the **for** statement is exhausted without jumping to the label FOUND, then the element $x$ was not in the array $a$. The average number of comparisons to find a matching element which may be at any position in the array with equal probability is $(n+1)/2$ but if the element is not in the array, no match is found and $n$ comparisons are made.

If the elements of the array $a$ are ordered, for example, into ascending alphabetical or numerical order so that we have (with an obvious interpretation if the $a_i$ are non-numerical) $a_1 \leq a_2 \leq \ldots \leq a_n$ then the algorithm can be improved by writing

```
for i:= 1 step 1 until n do
if x < a[i] then goto NOT FOUND else if x = a[i] then
                                          goto FOUND;
```

The average number of comparisons is now always $(n+1)/2$ whether the unknown $x$ is in the list or not.

If the relative frequency with which the different elements will be sought is known, and we can arrange the list as we wish, an improved directed scan is available. With the list $a_1, a_2, \ldots a_n$ and if the probability that the ith item has to be searched is $p_i$ we order the list according to decreasing probability. Thus $p_i \geq p_j$ if $i < j$. If this is done the average number of comparisons is

$$\sum_{i=1}^{n-1} ip_i + (n-1)p_n$$

when it is known that the item sought is present.

Directed scanning is usually adopted in cases when the comparisons between the keys and the unknown key give no further information as to where to find the unknown key in the list (apart from the obvious information that it is or is not the key being currently tested). Searching auxiliary storage devices is one application area in which these types of scanning algorithms are useful.

### 7.2.2 Controlled Scanning

In controlled scanning methods, the previous comparisons determine the subsequent ones. An effective controlled scan will be such that the information obtained by each comparison will be used to maximum effect in choosing the next element so that we minimize the number of elements to be scanned in later stages. Since comparisons, as we have noticed before, yield binary decisions, the most important controlled scan is naturally a binary search algorithm. The method is, in essence, the same as the Bisection Method used in numerical analysis for finding the root of an algebraic equation by progressively narrowing the interval in which the root is known to lie.

**Binary Search.** The method of binary searching aims to reduce to about half the number of elements to be searched at each key comparison; a comparison with the unknown element determines which half of the elements to search next, these elements are now halved and another comparison is made and so on. If the number of elements, n, is exactly $2^p-1$ (where p is an integer) there are no difficulties. However, in general, the number of elements will not be so convenient and our binary search algorithms must take this into account. In the detail of such an algorithm we have to decide whether to

(i) make approximate divisions at each stage (either by rounding up or rounding down), or

(ii) find the smallest integer p such that $n < 2^p$ and consider the list of length $2^p-1$.

(i) If we adopt the first strategy, consider the example of a list with $n = 13$ elements; rounding down will give successive numbers of elements in the 'halves' 6, 3, 1, while rounding up will give 7, 4, 2, 1.

If our search leads to the first value in the list, we must be careful because rounding down gives us $13 - 6 - 3 - 1 = 3$ and we have 'undershot' as we require the value 1. Whilst rounding up gives $13 - 7 - 4 - 2 - 1 = -1$ and we have 'overshot' and obtained an element not in the list. There are several methods of avoiding these troubles and a typical binary search algorithm of this kind is:

```
procedure binary search (key, n, x, FOUND, NOT FOUND);
        value n, x; integer n, x; integer array key;
        label FOUND, NOT FOUND;
        comment the keys are in the array key of size key [1:n].
        The unknown key we are searching for is x. The procedure
        exits to the label FOUND when the key is found and if it
        is not then the procedure exits to the label NOT FOUND;
        begin integer i, upper, lower;
                lower:= 1; upper:= n + 1;
        LOOP: i:= (upper + lower) ÷ 2;
                if key [i] = x then goto FOUND;
```

```
if key [i] < x then begin if upper = i + 1 then goto
                                        NOT FOUND;
               lower:= i + 1; goto LOOP
             end
        else begin if lower = i then goto NOT
                                        FOUND;
               upper:= i; goto LOOP
             end
end of binary search;
```

(ii) The second binary search algorithm uses the strategy of theoretically extending the array from size n to size $2^p-1$, where p is the smallest integer such that $n < 2^p$. The difficulty in this algorithm is that we obviously do not want to extend the actual array to size $2^p-1$ and therefore we will be in trouble if we attempt to access an element between n and $2^p-1$. The algorithm given below is one way of doing this - simply stopping the access to elements in this prohibited range.

```
procedure b s (a, n, x, FOUND, NOT FOUND);
    value n, x; integer n, x; integer array a;
    label FOUND, NOT FOUND;
    begin integer i, q, interval;
        for i:=1, 2 × i while i ≤ n do q:= i;
        interval:= q;
    for interval := interval ÷ 2 while x ≠ a[q] do
    begin if interval = 0 then goto NOT FOUND;
        if x < a[q] then q := q - interval
        else if q + interval ≤ n then q := q + interval
    end;
    goto FOUND
  end bs;
```

**Example:** Suppose the keys are mnemonics for an assembler language and the ordering is lexicographical. Let the key vector be given by the three-letter combinations shown in the left column below.

| | |
|---|---|
| ADD | Add |
| BCT | Branch and count |
| DIV | Divide |
| END | Stop |
| EXR | Exclusive OR |
| MUL | Multiply |
| SLL | Shift left logical |
| STO | Store |
| SUB | Subtract |
| TRA | Transfer |

We can derive the following binary search tree

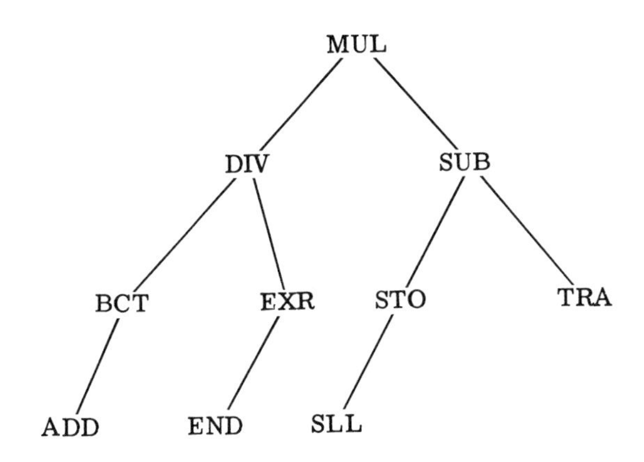


This binary search tree (BST) corresponds to the algorithm 'binary search' given earlier. There are several possible versions of the binary search algorithm and some of these would give slightly different binary search trees. The sequence of comparisons will always follow one of the paths of the above tree. The tree is a complete binary tree since there will be 1 node at level 1, 2 nodes at level 2, 4 nodes at level 3, ... $2^{k-1}$ nodes at level k until we reach the last level which will generally be only partially complete.

**Analysis of Binary Search Algorithm.** We have seen how the algorithm is equivalent to a binary search tree (BST) and we can use this idea together with the idea of the path length of the tree (see section 6.5) to derive the average number of comparisons per search in the binary search algorithm when all nodes are searched equally often.

The average number of comparisons, $\overline{c} = \sum_{i=1}^{n} \frac{l(i)}{n} = \frac{L}{n}$ where $l(i)$ is the level of node i in the BST and $L = \sum_{i=1}^{n} l(i)$.

Since this tree is complete, L is the sum of the first n terms of the series 1, 2, 2, 3, 3, 3, 3, 4, 4, 4, 4, 4, 4, 4, 4, 5, 5, ....

If $n = 1 + 2 + 2^2 + 2^3 + \ldots + 2^{k-1} = 2^k - 1$ then all the nodes at the lowest level of the tree are used and

$$L = 1 + 2.2 + 2^2.3 + 2^3.4 + \ldots + 2^{k-1}k \tag{7.1}$$

Consider $S(x) = 1 + 2x + (2x)^2 + \ldots + (2x)^{k-1} = \frac{2^k x^k - 1}{2x-1}$, (7.2)

$$(S - 1)x = 2x^2 + 2^2x^3 + \ldots + 2^{k-1}x^k.$$

Differentiating this series term by term we obtain

$$\frac{d}{dx}[(S - 1)x] = 2.2x + 2^2.3x^2 + \ldots + 2^{k-1}.kx^{k-1}.$$

If we put $x = 1$ in the above series and add 1 we have the series given in (7.1); therefore

$$L = 1 + \frac{d}{dx}[(S - 1)x]_{x=1}$$

Using (7.2)

$$L = 1 + 2^k(k - 1). \tag{7.3}$$

Therefore

$$\overline{c} = \frac{L}{n} = \left(\frac{n + 1}{n}\right) \log_2(n + 1) - 1 \tag{7.4}$$

since

$$k = \log_2(n + 1).$$

If however $n \neq 2^k - 1$ then the lowest level of the BST is incomplete. Taking $2^{k-1} \leq n \leq 2^k - 2$ we have to subtract from L the value $(2^k - 1 - n)k$ and thus (7.3) becomes

$$L = 1 - 2^k + k + kn .$$

Therefore $\bar{c} = \frac{L}{n} = \frac{1 - 2^k + k + kn}{n}$ where $k = 1 + \lfloor \log_2 n \rfloor$, where $\lfloor F \rfloor$ indicates the floor function and is the largest integer less than or equal to F. Thus

$$\bar{c} = \frac{1}{n}[(n + 1)\lfloor \log_2 n \rfloor - 2^{\lfloor \log_2 n \rfloor + 1} + 2 + n] \tag{7.5}$$

Thus for binary search algorithms the average number of comparisons is proportional to $\log_2 n$ as n increases, a considerable improvement on the value $n/2$ obtained for directed scanning algorithms.

The algorithms given for binary search are less effective if insertion of absent items in the list is also required. Serial searching with linkage is slow in the search part but insertion can be done very efficiently. When searching with insertion is required, however, the following binary tree structure has many advantages.

### 7.2.3 Binary Sequence Search Tree

If data is stored in a binary tree structure, which has the property that the left subtree of any node contains only keys numerically less than the key of that node while the right subtree contains only keys which are greater, a convenient controlled scanning method can be analyzed and algorithms for various necessary operations developed.

Given a list of keys the tree is constructed so that the first item is the root of the tree. Subsequent items are placed on the left of a node if its key is less than the key of the node and on the right of the node if the key is greater than that of the node. We place the item at the first unoccupied node.

**Example:** Given the keys

8, 1, 5, 9, 7, 12, 14, 6, 10, 4, 11

we would construct the following binary sequence search tree

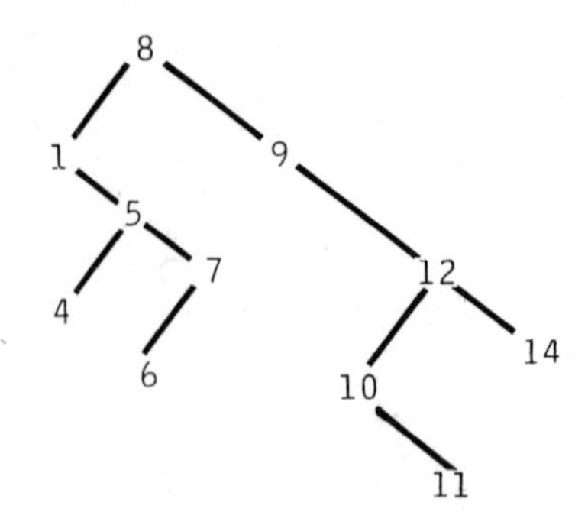


In order to illustrate possible searching, insertion and deletion algorithms for the binary sequence search tree we will consider each node to consist of three fields - the key, the left link and the right link. For the ith node these will be denoted by key [i], llink [i], and rlink [i] respectively.

**Search and Insert Algorithm.** Hibbard in 1962 wrote separate algorithms for searching and inserting, but it seems simpler to consider the two together. Thus, we wish to search the tree for a key value x and if it is not present we insert an item with this key in the appropriate position in the tree.

Let the address of the root of the binary tree be AR and let the value 0 be used to indicate a null link. We suppose there is a list A of available space and that S points to the first item in that list, so when we insert a new node we use the first item in the A list and move the A list pointer S down to the next member A[S].

The algorithm is

```
begin       integer i;
            i:= AR;
            while x ≠ key[i] do
            if x < key [i] then begin if llink [i] = 0 then
                                      begin llink [i]:= S;
                                            goto INSERT
                                      end;
```

```
                        i:= llink [i]
                end
        else begin if rlink [i] = 0 then
                        begin rlink [i]:= S;
                              goto INSERT
                        end;
                        i:= rlink [i]
                end;
    goto  FOUND;
INSERT:  key [S]:= x;  rlink [S]:= llink [S]:= 0;
         S:= A[S];
end search and insert algorithm.
```

Taking the example of a binary sequence search tree above, suppose we search for the key 2. When this is not found it will be inserted as the left subtree of key 4.

**Deletion Algorithm.** A simple algorithm which deletes the unwanted node (X) and preserves the basic properties of the BSST is

(1) If X is a leaf node delete it

(2) If X is a branch node and it has a null right subtree then replace X by its left subtree

(3) Otherwise replace X by its right subtree and put the left subtree of X as the left subtree of the smallest (leftmost) node of the right subtree of X.

In the example above, if we delete node 5 we get

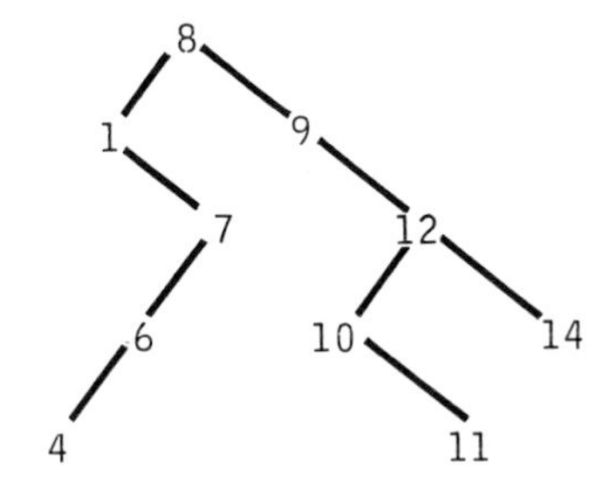

An alternative deletion algorithm was given by Hibbard and it has since been slightly improved by Knuth. The first two steps of this algorithm are the same as before but step (3) becomes

(3) Search the right subtree of X for the smallest (leftmost) node. Delete this node from its present position (replacing it by its right subtree if necessary) and put it in place of node X.

If we use the Hibbard-Knuth algorithm to delete node 5 in our example we obtain

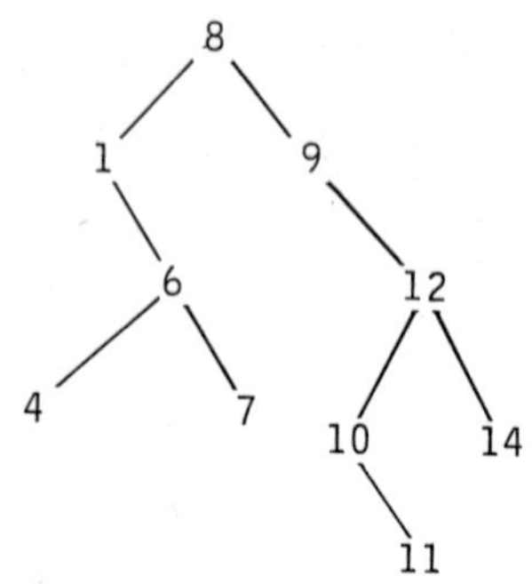


The Hibbard-Knuth deletion algorithm is better because it does not alter the expected average length of the BSST, whilst the simple deletion algorithm, if used frequently, will increase the average length of the BSST. The example above shows a case where the Hibbard-Knuth algorithm gives a better result.

The detailed deletion algorithms based on the representation given above are left as examples for the reader. We must when searching for the node X to be deleted remember the address of the father since the linkage of the father node is altered by the deletion algorithms.

The average length of the search path depends upon whether or not the item is in the tree and also upon the shape of the tree structure. Apart from those trees which have a very simple shape because, for example, they have been built up from items which occur in order, nearly all the analytical results that have been derived refer to trees formed from randomly arranged data - and calculated averages over all possible orders of the data of the tree. Under the latter assumption, which we feel would need some justification in most applications, the mean path length of a search for items present in a tree of n nodes or those absent is not far from $2 \log_e n$.

The above description of binary sequence search trees refers to those applications in which the tree is fixed throughout; an interesting modification which can increase the efficiency of searching by restructuring the tree to take account of the frequencies of accessing items has been developed by Coffman.

## 7.3 KEY TRANSFORMATION (SCATTER STORAGE) TECHNIQUES

In these methods, which are very useful when access has to be made to entries in a table in unpredictable order, the storage location is calculated from the key itself. A commonly occurring example in computer science is the assembler or compiler symbol table. In this case the names which appear during compilation are given a value which is the value of a current storage location and this value has to be consulted whenever the name is encountered in the program. Thus, the values are accessed in an unpredictable order. The table could be kept in alphabetical order and a binary search technique used but as we have seen this requires an average of $\log_2 n$ comparisons (where n is the size of the table). We will see that scatter storage techniques can do better than this. The sorted table is even worse if items are continually being added to the table, since either we must sort after each new item or wait until all the items have arrived, thus searching an unsorted table.

Let there be N items $(K_i, D_i)$ where $K_i$ is the key and $D_i$ the data for the ith item $(i = 1, 2, 3, \ldots N)$. Suppose that each key, K, is used to calculate an address $A = f(K)$ using the transformation function, f. One of the simplest cases is given by

$$f(K_i) = a\,K_i + b \qquad \text{where } a, b \text{ are constants.}$$

An application could be the storage of a table of 90 sines, cosines and tangents, each of which required one location, for the arguments 0, 1, 2, ... 89, so that we need three locations for each argument and thus take $a = 3$ while b is the address of the start of the table i.e. for sin 0. In this simple case the technique is known as table lookup.

The more interesting case is when the key transformation function

$f$ is not one-to-one. This is a very common case particularly with alphabetic keys because the amount of table storage required to give a one-to-one transformation is unacceptably large. If the key can be any five letter word then there are $26^5$ possible combinations and naturally the actual keys would only be a small subset of these, sparsely distributed over the large set of possible values. The problem, therefore, is to find a good transformation of the key which spreads the calculated addresses (usually called the hash addresses) uniformly over the available storage. If the hash address for a particular key is already filled by some other key and its data then we require some method of resolving this collision of keys - an operation which will entail checking the value of the key. Thus when the key transformation function is not one-to-one the key must be stored with the corresponding item of data.

### 7.3.1 Hash Addressing

Most hash addressing schemes use transformation functions which produce a certain number $m$ of bits and in these cases it is convenient to use the $m$ bits as index to a table of size $n = 2^m$. The possibility of collisions means that there must be a method of distinguishing empty locations in the table and of initializing all the storage. For example, if a zero key is not included in the actual keys, it can be used to recognize empty locations. We should like key transformations which distribute the actual keys into addresses fairly uniformly over the available storage. Most of the hash addressing methods used are empirical and have been used because they are fast and effective in practice. Some of these empirical methods are discussed below in detail.

(1) Regard the key $K_i$ as a binary integer, form $K_i^2$ and select $m$ bits from the middle to address the $2^m$ stores allocated. This method is particularly useful if the key is a single word. The middle digits of a square depend on all the bits of moderate-sized keys and thus different keys are likely to give different hash addresses even if the keys have common bit patterns, for example at the beginning or the end.

(2) If $m$ bits are required for the hash address, cut the key $K_i$ into sections $m$ bits long, add them together and use the $m$ least significant bits of the sum as the hash address. This method can be used with single or multiword keys.

(3) If we have multiword keys, then we can use m bits taken from the product of the words. We must clearly be careful in cases where the key contains blanks which are internally coded as zero and when partial word keys are padded out with zeros as such cases lead to a large number of the calculated addresses being zero.

(4) Integer divide $K_i$ by the size of storage allocated and use the remainder as the hash address. Unless there is a known freakish distribution of the $K_i$ clustering is likely to be less pronounced if the size of the table n is chosen to make n a prime. We can of course use only some of the bits of the remainder as the hash address. This method is liable to give the same hash address very often if there are definite patterns in the keys or if the keys are filled with blanks.

(5) Truncate to the m most significant bits of the key. This method is very simple but will give trouble in cases where there are common features at the beginning of the keys.

### 7.3.2 Collisions

When the scatter storage table is partially filled, any newly computed hash addresses may cause a collision with a hash address computed for a previous key. Another place must be found for the new key and some procedure entered which calculates new addresses until an empty storage location is obtained. This same procedure must be used when the key must be located in a search. In practice it is not necessary to distinguish between the insertion algorithm and the search algorithm for a scatter storage table. The same process is carried out for both until either an empty location is reached or the desired key is found. Since several keys can generate the same hash address, the key must be stored in the scatter storage table while the data associated with the key may follow it there. It may, however, be more convenient to store a pointer after the key which points to the appropriate item in storage; this avoids moving items of data in store, makes deletion relatively easy, and caters straightforwardly with items of data of varying length which would otherwise either considerably complicate the storage in the table or waste space by allocating every item as much space as the largest item.

**Overflow Tables.** When a collision occurs we must either find an alternative slot in the scatter storage table or have an 'overflow' table. If we store in the scatter storage table, then the method of handling collisions must be able to probe every space in the table, and it will also mean that the number of collisions will rise rapidly as the table becomes nearly full. The idea of an 'overflow' table for storing keys after the occurrence of a collision has some attractions, particularly when combined with a chaining method which will be discussed further in the section on chaining. The simplest idea is just to put the overflowed item in an overflow table. Then the search algorithm after making the key transformation sees if the item is in the scatter storage table and if not, performs a directed scan of the overflow table. The methods that use just the scatter storage table use in effect the empty locations in that table as the overflow table. We now look at some methods of handling collisions in this situation where the whole efficiency of a scatter storage technique will largely depend on the efficiency of its collision handling procedures.

**(a) Linear.** In this method if a collision occurs, we advance $s$ storage locations from the calculated address. The scatter storage table is considered circular, so that location $n$ is followed by location 1. We continue advancing $s$ locations at a time until a free storage location occurs if we are inserting or until we find the key sought if we are searching. If $s = 1$ then we are doing a simple forward scan until we locate the key or an empty location. This is a so-called open addressing technique.

The main disadvantage of this method is that it often produces clustering of the occupied locations. Of the methods described for handling collisions, this in practice is the least effective in terms of the average number of probes necessary to retrieve an item. It can be seen that the second collision at an address $L$ would cause an attempt to store at $L + 2s$, which is the place where the first collision at $L + s$ (more likely because of the first collision at $L$) will be stored; and so on. The drawback is that several sequences of collisions are stored in the same sequence of locations - phenomenon called 'primary clustering'. The method is, of course, somewhat simpler to program than other methods and the amount of calculation done at each stage is very small.

The procedure for deleting items is not simply to mark the location empty. If this were done, then there may be items that had collided with the deleted item and they would now be unreachable. A complex series of moves to close the gap could be made, but the simplest method is to have a special marker for deleted locations, in much the same way that there is a special marker for empty locations. When searching, we skip past locations marked as deleted at the cost of accessing the location and checking its contents and when inserting, we can insert in either a deleted or empty location. The disadvantage of this method is that space only is saved while searching time is not reduced.

(b) **Quadratic.** This method is similar to the linear method except that after each successive collision for a given key the number of locations advanced is increased by 1. Thus after $i$ successive collisions starting at the hashing address, locn, we try $locn + 1 + 2 + \ldots + i = locn + i(i + 1)/2$ (modulo the table size). If the table size is $2^m$ the whole table will be searched but for a general size it will not - for example, if the table size is prime only half is searched.

The main advantage of this method over the linear method is that it eliminates the worst clustering. This is because if the quadratic search goes through locations holding the keys $K_1, K_2, \ldots$ to get to $K_j$ in the scatter storage table, it will not go through this chain if the quadratic search starts at $K_2$ or $K_3$ etc. This elimination of the primary clustering makes the quadratic method better than the linear method. However, another kind of clustering, called 'secondary' clustering, remains. This is caused by the fact that two keys which hash initially to the same location, trace the same sequence of locations; that is those keys with the same hash address form a cluster. A variation of the quadratic method (called the Quadratic Quotient Method) which eliminates this secondary clustering has recently been suggested. This method simply lets the quadratic term vary with the key $K$ instead of being a constant.

The deletion algorithm for the quadratic method is similar to that for the linear case.

**(c) Random.** This method of handling collisions works as follows.

(i) First perform the key transformation on the key K:

$l = f(K)$ .

(ii) If the item being searched for is at this location or if this location is empty the algorithm is completed.

(iii) If some other item is at the location $l$ a collision has occurred and we call a pseudo-random number generator to get an integer offset $r_i$. Make the next comparison at location $l + r_i$ (with the table wrapped round end to beginning) and go to step (ii).

The pseudo-random number generator should be a simple one and have the property that it generates every integer from 1 to $n - 1$ (where $n$ is the table size) once and only once while less emphasis can be placed on the statistical properties of the sequence. A pseudo-random number generator which satisfies these requirements for a table size $n = 2^m$ is

(1) Initialize $r := 5$ every time a search or insertion operation is commenced.

(2) Set $r := r \times 5$;

(3) Mask out all but the lower $m + 2$ bits and store the result in $r$.

(4) Final offset $r_i := r \div 4$ (where $\div$ is integer divide).

Use this value of $r_i$ to find the next location and if there is another collision repeat steps (2), (3) and (4) of the pseudo-random number generator.

The deletion algorithm is similar to that for the linear case.

**(d) Chaining.** There are two variants of this basic method usually known as direct chaining and indirect chaining. In both the methods we use some extra storage space for links in order to get a more efficient method of handling collisions. Thus all the items which generate the same hash address are to be found on a linked list (or chain) starting at that hash address. The link in the last item on a chain should be a special marker that indicates there are no more items on this chain. As before such a special marker would be a number which cannot be taken for a storage address.

The searching and insertion algorithm for direct chaining is as follows:

(1) Compute the hash address by the key transformation.

(2) If the address is empty insert the item here giving it the special marker as link.

(3) If the address is occupied search the chain starting from this address until either the key is found in which case the search is ended or the key is not found.

(4) If, when the key is not found, the hash address is occupied by the head of the chain then find a new address in the scatter table and insert the new item at this address. Link the new address into the chain preferably near the head of the chain in order to avoid unnecessary probes down the chain.

(5) If the hash address is occupied by an item which is not the head of a chain (it will be, therefore, an item which is not at its own hash address) then the old item must be moved to some new unoccupied address and the new item inserted in its place. Moving the old item necessitates finding an empty location and altering the linkage on its chain.

It is this last step in the algorithm which is its main disadvantage as it requires moving items in storage and the extra programming to do this and alter the linkages. An advantage of the method is that we can still use the same searching and insertion strategies when overflow occurs in the scatter storage table.

The second variant of the basic chaining method is called indirect chaining or overflow with chaining. In this method we treat every item that collides at any hash address as an overflow item which is stored in an overflow table with the other overflow items. This has the advantage over direct chaining that items need never be moved in storage once they are entered, but it does, of course, use extra storage space since items are put in the overflow table before the scatter storage table is full. If the scatter table is kept small this is not too serious a drawback.

The deletion algorithm for these chaining methods is:

(1) If the item is not the chain head, link the chain round this item and mark the address empty.

(2) If the item is the chain head but there is only one item on the chain mark the address empty.

(3) If the item is the chain head and there is a chain starting from it then it can either be marked as deleted or one of the later items in the chain moved up to be the chain head and the appropriate changes in the linkage made.

### 7.3.3 Comparison of the Methods of Handling Collisions

The efficiency of these methods is best expressed in terms of the average number E of probes necessary to access a randomly chosen item in the scatter storage table. For any item the number of probes to access the item is the same as the number of probes necessary to insert the item in the table initially, except in those chaining methods where items are moved. The number E depends on $\alpha$, the proportion of the table that is occupied:

$$\alpha = \frac{\text{number of items in the table}}{\text{size of the table (n)}} .$$

The mathematical analyses of the various methods involve certain approximations in addition to the assumptions about randomness which themselves may be at best approximate in practice. One of the simpler analyses which is suitable for the quadratic quotient method and is approximately correct for the random method is outlined here as an example of the arguments that are needed.

When j items have been inserted in the table, the probability that just one probe is needed is $(n - j)/n$, the probability that a vacant location is found at once.

The probability that exactly two probes are needed is the product of the probability that the first location tried is full $(j/n)$ and the probability that the second one tried in the remaining $(n - 1)$ is free $(n - j)/(n - 1)$. Similarly the probability that exactly i probes are needed is

$$\frac{j}{n} \cdot \frac{j-1}{n-1} \cdot \frac{j-2}{n-2} \cdots \frac{j-i+2}{n-i+2} \cdot \frac{n-j}{n-i+1} .$$

Thus the expected number of probes to insert $(j + 1)$st item is

$$e_{j+1} = \frac{n-j}{n} + 2 \cdot \frac{j}{n} \cdot \frac{n-j}{n-1} + \ldots + (j+1) \cdot \frac{j}{n} \cdot \frac{j-1}{n-1} \cdot \frac{j-2}{n-2} \cdots \frac{1}{n-j+1} .$$

By induction we can verify that

$$e_{j+1} = 1 + \frac{j}{n-j+1} .$$

Assuming each of the k items in the table are equally likely to be searched for

$$E = \sum_{j=0}^{k-1} \frac{1}{k} e_{j+1} = \frac{1}{k} \sum_{j=1}^{k-1} \left(\frac{n+1}{n-j+1}\right) = \frac{n+1}{k} (H_{n+1} - H_{n-k+1}) , \qquad (7.6)$$

where

$$H_n = \sum_{i=1}^{n} \frac{1}{i} = \log_e n + \gamma .$$

If we take $\alpha \doteqdot \frac{k}{n+1}$ , then (7.6) becomes

$$E \doteqdot -\frac{1}{\alpha} \log (1 - \alpha) \qquad (7.7)$$

which gives the average number of probes required.

The references cited later give other examples of the analysis of collision handling methods under various assumptions; in this section we will only quote some appropriate approximate formulae and display a numerical comparison of the results.

For the linear method

$$E \doteqdot (1 - \tfrac{\alpha}{2})/(1 - \alpha) ;$$

for the random and quadratic methods

$$E \doteqdot -\frac{\log(1-\alpha)}{\alpha} + \frac{1}{2\alpha} \left\{\frac{1-\alpha^2}{2} \log(1-\alpha) + \alpha - \frac{1}{4} + \frac{(1-\alpha)^2}{4} \right\} ;$$

for the quadratic quotient method

$$E \doteqdot - [\log(1 - \alpha)]/\alpha ;$$

and for the chaining method

$$E \doteqdot 1 + \frac{\alpha}{2} .$$

A comparison of the methods is given in the table.

Although a first glance at the table suggests that the chaining method is markedly superior to the other methods, it must be remembered that it uses more store for its overflow items and so is not directly comparable. It is, of course, the only method which can accept loadings with $\alpha > 1$.

| Loading | Average number of probes, E | | | |
|---|---|---|---|---|
| | Linear method | Random and Quadratic methods | Quadratic quotient method | Chaining method |
| 0.1 | 1.06 | 1.06 | 1.05 | 1.05 |
| 0.5 | 1.50 | 1.44 | 1.39 | 1.25 |
| 0.75 | 2.50 | 1.99 | 1.85 | 1.38 |
| 0.9 | 5.50 | 2.80 | 2.56 | 1.45 |
| 0.95 | 10.5 | 3.45 | 3.15 | 1.48 |
| 1.00 | $\infty$ | $\infty$* | $\infty$ | 1.5 |
| 2.00 | - | - | - | 2 |

* As $\alpha \to 1$ the difference caused by secondary clustering $\to 0.375$.

### 7.3.4 Applications of Scatter Storage Techniques

(1) **Symbol Tables.** This is one of the most obvious applications of scatter storage techniques. One may be decoding the mnemonic codes for assemblers to get the correct bit pattern of the order, or decoding a macro name to obtain the address holding its first instruction. A similar type of application is an identifier table used by a compiler or assembler. In these applications if an error occurs we have to unhash in order to print out meaningful diagnostic messages.

(2) **Sparse Arrays.** As we have seen in chapter 4 the array subscripts uniquely specify a storage location for an item. This is an obvious way to store and access array items in dense arrays. However, if the array has only a few items (e.g. a sparse matrix) we can use the subscripts as a key and from them compute a hash address. The array item

is accessed through this hash address. The subscripts (which are in this case the key) must be stored with the item itself in the scatter storage table. This method of storing sparse arrays has many advantages when the array items are accessed in an unpredictable order but it is not such an effective method if systematic operations such as addition, subtraction and multiplication are needed.

## 7.4 BIBLIOGRAPHY

Elementary introductions to searching which cover simple scanning and binary searching are given in Arden [1] and Galler [2]. A more complete coverage of searching which includes scanning and key transformation techniques is given in Iverson [3]. However, this book is not easy to read and all the algorithms are in Iverson's complete APL notation. A good up-to-date survey of table searching (for static tables only) has been given by Price [4].

Section 7.2.3 on binary sequence search trees was based on Hibbard [5], a paper which reads well and still contains much useful information. The modification of the tree for frequently accessed items can be found in Coffman and Bruno [6].

Scatter storage techniques (section 7.3) have been excellently summarized by Morris [7]. This paper also extends the ideas to scatter index tables, virtual scatter tables, and scatter tables on paged machines. For the individual collision handling techniques, Schay and Spruth [8] covers the linear method, Maurer [9] the quadratic method, Bell [10] the quadratic and the quadratic quotient method, McIlroy [11] and Morris [7] the random method, and Johnson [12] the chaining method.

Since the first edition of this book the third volume of Knuth's The Art of Computer Programming has been published [13]. In Volume III there is a very comprehensive study of searching methods with sections on Sequential Searching, Searching by comparison of keys, Digital Searching, Hashing, and Retrieval on Secondary Keys. It is an ideal starting point for any reader requiring further information on the topics covered in this chapter.

Three recent articles in the ACM Computing Surveys Journal have covered Searching methods. The first by Severance [14] is a survey and generalised model based on a TRIE-TREE search mechanism, the second by Nievergelt [15] covers binary sequence search trees, and the third by Maurer and Lewis [16] covers key transformation techniques.


[1] B. W. Arden: An Introduction to Digital Computing, pp. 284-91. Addison-Wesley, 1963.

[2] B. A. Galler: The Language of Computers, pp. 97-104. McGraw-Hill, 1962.

[3] K. E. Iverson: A Programming Language, pp. 133-58. John Wiley and Sons, 1962.

[4] C. E. Price: Table Lookup Techniques, ACM Computing Surveys vol. 3, no. 2 (June 1971), pp. 49-65.

[5] T. N. Hibbard: Some Combinatorial Properties of Certain Trees with Applications to Searching and Sorting. J. A. C. M. vol. 9 (1962), pp. 13-28.

[6] E. G. Coffman and J. Bruno: On File Structuring for non-uniform Access Frequencies, B. I. T., vol. 10 (1970), pp. 443-56 (also University of Newcastle upon Tyne Computing Laboratory Technical Report 6, January 1970); Optimal Binary Search Trees, Procs. IFIP Congress, Ljubljana, Yugoslavia, August 1971.

[7] R. Morris: Scatter Storage Techniques, C. A. C. M. vol. 11, no. 1, January 1968, pp. 38-43.

[8] G. Schay, Jr. and W. G. Spruth: Analysis of a File Addressing Method, C. A. C. M. vol. 5, no. 8 (August 1962), pp. 459-62.

[9] W. D. Maurer: An Improved Hash Code for Scatter Storage, C. A. C. M., vol. 11, no. 1 (January 1968), pp. 35-8.

[10] J. R. Bell: The Quadratic Quotient Method: A Hash Code Eliminating Secondary Clustering, C. A. C. M. vol. 13, no. 2 (February 1970), pp. 107-9.

[11] M. D. McIlroy: A Variant Method of File Searching, C. A. C. M. vol. 6, no. 3 (March 1963), p. 101.

[12] L. R. Johnson: Indirect Chaining Method of Addressing on Secondary Keys, C. A. C. M. vol. 4, no. 5 (May 1961), pp. 218-22.

[13] D. E. Knuth: The Art of Computer Programming, Volume 3: Sorting and Searching, pp. 389-570, Addison-Wesley, 1973.

[14] D. G. Severance: Identifier Search Mechanisms: A Survey and Generalised Model, ACM Computing Surveys, Vol. 6, No. 3, (September 1974), pp. 175-94.

[15] J. Nievergelt: Binary Search Trees and File Organization, ACM Computing Surveys, Vol. 6, No. 3, (September 1974), pp. 195-207.

[16] W. D. Maurer and T. G. Lewis: Hash Table Methods, ACM Computing Surveys, Vol. 7, No. 1, (March 1975), pp. 5-20.

## EXAMPLES 7

[7.1] For one of the two binary search algorithms write a suitable insert algorithm and deletion algorithm.

[7.2] Given the keys

MUL, STO, DIV, SUB, EXR, TRA, SLL, ADD, END, BCT ,

in that order on an input stream, construct the binary sequence search tree. How would this tree be altered if

(a) the key JRZ were added;

(b) the key STO were deleted?

[7.3] Given four keys in all possible orders how many different binary sequence search trees are there? Show how to obtain this result for n keys from a knowledge of the results for less than n keys.

[7.4] In the random method of handling collisions in a scatter storage table (see section 7.3.2(c)) what are the offsets $r_i$ produced by the pseudo-random number generator if the table size is $8 = 2^3$?

[7.5] Write detailed algorithms for the two methods of deleting in a binary sequence search tree. The two deletion algorithms are explained generally in section 7.2.3.

[7.6] The analysis of the Binary Search Tree given in section 7.2.2 uses in effect generating functions. Considering the case with exactly $2^k - 1$ nodes give an alternative proof of equation (7.3) using recurrence relations.

Hint: If $u_k$ is the total number of comparisons for a tree with $2^k - 1$ nodes, show the basic recurrence relation is

$$u_k = 2u_{k-1} + 2^k - 1.$$

[7.7] Suppose we have a scatter table whose corresponding hash code is a single bit. Let N be the size of the scatter table and suppose we store keys with a hash code of 0 successively into (relative) locations 1, 2, 3, ... and keys with a hash code of 1 into locations N, N - 1, ... .

Thus, to search for a given key in the table a search algorithm first generates the hash bit and then performs a directed scan from bottom up if the hash bit is 0 and from top down if the hash bit is 1. We may assume that there are two counters, maintained so as to indicate the number of keys with 0 hash bits and the number of keys with 1 hash bits stored in the table.

(a) Suppose the number of keys stored in the scatter table varies with time. Fully describe insertion and deletion algorithms so that the search algorithm above will always find a desired key.

(b) Suppose the hash function used with the table is biased so that with fixed probability $p$ an arbitrary key has a hash code of 0 and with probability $q = (1 - p)$ a hash code of 1. Assuming each key in the table is equally likely to be accessed show that the average number of comparisons to locate a key known to be in the table is approximately

$$\overline{t} = (n/2)[1 - 2pq]$$

where $n \leq N$ is the number of keys stored in the table.

(Newcastle 1970)

[7.8] A variant of the binary search makes use of a three-way comparison, i.e. a test is made whether the item at the mid-point is less than, equal to, or greater than the item being sought. Show that the maximum number of tests required to locate an item in a table of $n$ items is $m + 1$, where $m$ is the greatest integer not exceeding $\log_2 n$, and that the average number of tests required exceeds $m - 1$.

[You may assume the result $\sum_{k=0}^{n} (k + 1)2^k = 1 + n2^{n+1}$.]

(Southampton, Diploma 1969)

[7.9] Explain how a binary sequence search tree (BSST) is built, and draw such a BSST for the following sequence of keys

8, 5, 1, 9, 7, 18, 15, 6, 3, 2, 10, 4.

Describe how you search for a key using a BSST and how you delete a key no longer required. How would the above BSST alter if the node 5 was deleted?

If $M^*$ is the mean path length for an item in the tree structure and $\overline{M}$ is the mean path length for an item not in the tree structure show that for $n$ items

$$M^*(n) = \frac{2(n+1)}{n} H_n - 3$$

where

$$H_n = \sum_{k=1}^{n} \frac{1}{k} .$$

You may assume that for $n$ items the following recurrence relation holds for $\overline{M}(n)$

$$\overline{M}(n) = \frac{2}{(n+1)} + \overline{M}(n - 1) .$$

If there are ten different keys, compare the search time (or mean path length) for the best possible BSST and the worst possible BSST. Assuming the keys are 1, 2, ..., 10, give an input sequence for the keys which will give a best possible BSST.

(Newcastle 1971)

[7.10] A large table is to be mapped into the core storage of a computer. To achieve fast search times a hash table is to be used. However, it is necessary to be able to output the items of the table in order of the keys, and for this purpose the items are to be linked as a binary tree. Each node of the tree is to contain a table item. The left subtree of a node containing key $k_i$ will be either null or a tree of nodes containing keys $k_j < k_i$. Similarly, the right subtree will be either null or a tree of nodes containing keys $k_j > k_i$.

In operation a key is to be presented to the hash mechanism to search the table. If the key is found in the table, information relating to the key is retrieved. If not found, the key and related information are to be added to the table and incorporated in the tree. Initially, the table will be empty. The table may be output at any time.

(a) Describe a possible internal storage structure for the table.

(b) Outline a method by which the table may be searched and if necessary added to.

(c) Outline a method by which the table may be output.

(Southampton, Diploma 1971)

[7.11] A direct access file F, is allocated B buckets of an exchangeable disc store. The file is to contain:

$n_0$ records of length $l$ words,
$n_1$ records of length $2l$ words,
$n_2$ records of length $4l$ words,
$n_3$ records of length $8l$ words.

The length of a bucket is $8l$ words and the number of records in the file satisfies

$$n_0 + 2n_1 + 4n_2 + 8n_3 \leq 8B .$$

Each record has a unique key and each size of record may be assumed to contain two spare words for administrative use.

Describe, with the aid of store map diagrams, how you would organize these records in the file F given that the following operations are equally likely:

(i) Retrieve the record whose key is k.

(ii) Add a new record of length $p.l$ with unique key k (p = 1, 2, 4 or 8).

(iii) Delete the record whose key is k.

(Essex 1971)

[7.12] Describe directed scanning as a method of searching a list of elements $a_1, a_2, \ldots, a_n$ to find an unknown element x. Consider the following three cases

(a) the list is unordered,

(b) the list is ordered,

(c) the list is ordered according to decreasing probability. In this case therefore the relative frequency with which different elements will be sought is known and the list is ordered such that $p_i \geq p_j$ if $i < j$, where $p_i$ is the probability that the search is for the $i^{th}$ element.

For each of these cases give the average number of comparisons required for a successful and unsuccessful search.

A table contains N identifiers out of a maximum possible of M. If an incoming identifier is not one of the N in the table we assume it can be stored in the next location in the table. Assuming that each of the M identifiers is equally likely, find the average number of comparisons to match an incoming identifier or store it in position N + 1.

Show that for M = 60 and N = 40 the average number of comparisons is 27.

(Newcastle 1974)

[7.13] Describe the method of searching an ordered list of keys $k_1, k_2, \ldots, k_n$ known as binary search. Give a detailed algorithm for your method and draw the binary search tree that would be obtained for the 24 keys

05, 07, 18, 19, 24, 27, 29, 30, 32, 37, 42, 45,
50, 54, 65, 68, 69, 75, 76, 81, 83, 85, 94, 97.

If the number of keys $n = 2^k - 1$, show that the average number of comparisons when using binary search is

$$\left(\frac{n+1}{n}\right)\log_2(n+1) - 1.$$

How would the further requirement that keys be added to the list dynamically, affect your algorithm?

(Newcastle 1973)

[7.14] A binary sequence search tree (BSST) is to be constructed in which the data value at any node is less than all those in its right subtree and greater than all those in the left.

The BSST is to be constructed using ALGOL W records and each node has four fields

(1) A REFERENCE which points to the left subtree (LLINK).

(2) A REFERENCE which points to the right subtree (RLINK).

(3) An INTEGER which is the data value at the node (DATA).

(4) An INTEGER which is a count of the number of times that particular data value has occurred. (COUNT.)

Write procedures which will (a) search for a particular item of data and update COUNT if it finds it or insert it into the BSST if it is not there.

(b) Find the node in the BSST with the largest value of COUNT.

(Newcastle 1975)

[7.15] (i) What is a hashing function, and what are the desirable properties sought in such functions? Exemplify briefly several basic types of hashing functions.

(ii) Briefly describe commonly used techniques for collision handling in hash table searches. Make reference to the performance of the methods at varying load factors.

(iii) Describe how address calculation techniques can be used for sorting sequences of keys. Outline how collisions are handled.

(Leeds 1975)

[7.16] Programming Project: Searching Using a Binary Tree

The purpose of this project is to first of all build the binary sequence search tree for the keys as they are encountered. However, if there is a large amount of searching to be done, we may find that the tree that was given by the original order of the keys is not the best. In order to improve it in the light of the new information (i. e. how often each node was searched for) only local changes in the tree structure are allowed. A complete rearrangement would be too time consuming. Such an algorithm is given in E. G. Coffman and J. Bruno 'On file structuring for non-uniform access frequencies', Newcastle University Technical Report 6, January 1970, and J. Bruno and E. G. Coffman 'Optimal Binary Search Trees' Procs. IFIP Congress, August 1971.

[7.17] Group Searching Project

In this group project, each student will have to develop his own searching method and program it in a suitable language. The group will be expected to cooperate on questions of evaluation of the searching methods, i.e. deciding on test sets of data. There are two subgroups, one examining hashing methods (4 projects) with fixed collision handling, and the other group examining collision handling methods (7 projects) with fixed hashing method.

Searching Methods: The general references are given in section 7.4, Bibliography. Section 6.4 in Knuth [13] is particularly useful and Morris [7] and Maurer and Lewis [16] are worth consulting.

Key Transformation Methods

(a) Hashing Methods: These methods are described in section 7.3.1 and in addition to the references given above there is a very useful paper by V. Y. Lum, P. S. T. Yuen and M. Dodd, Key-to-Address Transfer Techniques, CACM, Vol. 14, No. 4, (April 1971), pp. 228-39.

This subgroup is expected to look into hashing methods and for this purpose they should keep the collision handling method the same for all the different methods of hashing the keys, e.g. they could all use chaining with overflow but could choose another method if they preferred.

Some hashing methods to be considered are:

1. Division

   Divide the key by a positive integer - the remainder becomes the address in the SST (scatter storage table).

2. Mid-Square

   Square the key and take sufficient digits from the middle.

3. Folding

   Partition the key into parts the same length as the address length.

Then add them together. An alternative is to use an exclusive-or operation instead of adding the parts.

4. Lin's Method

See Lum, Yuen and Dodd, p. 230 for a description of this method.

(b) Collision Handling Methods: This subgroup is expected to examine different methods of handling the collision of keys in the scatter storage table (SST). In order to do this, they should choose a hashing method and keep it the same for all different collision handling methods.

Some collision handling methods to be examined are:

5. Linear Probing

If a collision occurs, move on a fixed number of places. (See section 7. 3. 2(a) and Morris [7].)

6. Quadratic

When a collision occurs the number of places we move forward is determined by a quadratic function as described in section 7. 3. 2(b). (See also Maurer [9].)

7. Random

This method is described in section 7. 3. 2(c). (See also Morris [7] and McIlroy [11].)

8. Quadratic Quotient

An improved variant of the Quadratic Method. (See Bell [10].)

9. Modified Linear

A method of improving the straightforward linear probing. (See R. P. Brent, A modified linear scatter storage technique, IBM Report, Report No. RC 3460, also CACM, Vol. 16, No. 3 (February 1973), p. 105.)

10. Direct Chaining

This puts colliding keys on a chain in the SST as described in section 7. 3. 2(d). (See also Morris [7].)

11. Indirect Chaining with Overflow

If keys collide then they are put on a chain but the chain goes over into an overflow area. The SST should therefore not be as big as in direct chaining. This method is also described in section 7. 3. 2(d). (See Morris [7] and Johnson [12].)

# 8 · Sorting

## 8.1 INTRODUCTION

Sorting can be described as arranging a set of keys in ascending or descending order. The purpose of sorting is to make the files of records easier to handle, in other words to improve the effectiveness of the search algorithms which are to be applied, either within a computer or externally. Thus sorting is closely related to, or even just a part, of searching. However, so much attention has been devoted to sorting that it is usually treated as a subject in its own right. Nevertheless it is important not to lose sight of the reasons for sorting.

### 8.1.1 Basic Definitions

A Record (or item) is the key plus the information (or data) content of the record.

A Key is a set of symbols (usually alphanumeric) which identify the record and can be used to determine the position of the record in an ordered sequence of records.

A File is a sequence of records - often ordered.

Thin keys are a set of keys in which few of the possible values actually occur.

Thick keys are a set of keys in which most of the possible values occur, e.g. if the range of possible keys is the numerical values 0-999,

(i) if only keys 0, 5, 721, 846, 947 appear then the set of keys are thin.

(ii) if all keys except the above five values appear then the set of keys are thick.

### 8.1.2 Detached Keys

In many applications the information content of the records is large. On the other hand the key itself is usually quite short, so that we can considerably improve the efficiency of our algorithms by detaching the key from the record and not subsequently moving the records during the progress of the algorithm. When the key is detached from the information part of the record, it must be augmented by the address of its associated information. The sorting process is carried out on the augmented keys and the information part of each record remains in the same place in the store. When the sorting process is finished we can reconnect the key and its information and place the records in order in the computer if required - for example when the file is to remain there permanently. In most circumstances it is easier to leave the records where they are and access them through the sorted augmented keys.

### 8.1.3 The Choice of Sorting Method

The sorting process can be considered to be in three stages:

(1) Reading the keys,

(2) Deducing the required place in the sequence,

(3) Moving the records into place.

Whilst any sorting process involves these three stages, different methods invoke them in varying ways. There are very many different sorting methods each with different characteristics and no one which is best in all circumstances. The main factors which effect the choice of the sorting method are:

(a) The type of computer storage available for the records. There are three principal categories:

(i) The store may allow rapid access, equally quickly to any of the records - fast, random access e.g. core stores.

(ii) Less rapid access may be possible, again random or nearly so; for example with magnetic drums or disc stores.

(iii) Access may be possible only to the records in a definite sequence as on a magnetic tape.

Ideas of speed change with advancing technology but in the early nineteen-seventies, fast access means in a micro-second or less while type (ii) devices need a few milliseconds for their access.

(b) The computer architecture as it appears to the user; the important effects are closely connected with the previous factor of computer storage, the way storage is connected and organized by both the hardware and the operating system. Relevant considerations are central processing speeds relative to store access times, whether the operating system is using paging or another storage allocation method and what specific algorithms are employed.

(c) The amount of data to be sorted. A very large volume of data will limit the type of storage that can be used, since only the slower backing store will be sufficiently large.

(d) The ordering of data. The performance of some sorting methods is considerably improved if the data are already partially ordered. It is therefore useful to know whether the data will be random, or partial ordered, or consist of several ordered subsequences.

(e) The thickness of the key. The distribution of the keys over the given range is a factor in deciding whether some sorting methods are applicable or not. For example, the simple pigeonhole sort is only effective with thick keys.

(f) The criteria by which efficiency is to be assessed. We can seek to minimize either

(i) The number of key comparisons,

or (ii) The number of record transfers (for example, by the use of detached keys),

or (iii) The extra storage space required.

However, it is more likely that rather than minimize one of these quantities we will seek some compromise involving all three.

These factors are not independent of each other; for example, we can often reduce the number of comparisons at the expense of extra storage. The use of detached keys reduces the number of record transfers, usually at cost of more storage. There is a voluminous literature on sorting; in this chapter we cover a few of the different sorting methods and discuss their strengths and weaknesses.

### 8.1.4 Minimum Number of Key Comparisons

In the analysis of sorting methods it is useful to know a lower limit on the number of key comparisons to indicate how well the various methods perform. Such a limit can be obtained from the basic concepts of information theory (introduced in chapter 2), where we noticed that one comparison, which tells us whether key [i] > key [j] or not, yielded one bit of information. According to information theory we need $\log_2 M$ bits of information to select one alternative out of M equal possibilities. Thus when we are sorting we select one particular ordering of the N keys out of all the N! possible orderings and this requires a minimum number of comparisons

$$\text{Min}_c = \log_2 N! \tag{8.1}$$

If we replace N! by Stirling's approximation we obtain

$$\text{Min}_c \doteqdot \log_2 \sqrt{(2\pi N)} + N\log_2 N - N\log_2 e \tag{8.2}$$

As N becomes large the right hand side of (8.2) is dominated by the second term. Thus the minimum number of simple key comparisons we can expect in an efficient sorting method is about $N \log_2 N$. We shall notice that several of the methods we discuss obtain more information from an examination of a key than the single bit of a key comparison.

## 8.2 INTERNAL METHODS

The methods described in this section are applicable to random-access storage only and are in general not convenient for data stored sequentially, for example on magnetic tape. It will, however, be noticed that nearly all the sorting methods described in this chapter will be acceptable to a large degree using random access storage although the converse is not true for sequential storage. In our discussion of the various methods it will be assumed (without loss of generality) that the sorting process is required to put the keys into ascending order, and that the keys are numeric. The sorting methods introduced will be evaluated by the criteria of section 8.1.3(f), together with their ability to benefit from ordered data.

### 8.2.1 Simple Sorting Methods using Selection, Exchanging and Insertion

**Selection** (sometimes called Linear Selection). The first method we will describe is conceptually one of the simplest methods of sorting the keys $k_1, k_2, \ldots k_n$ into ascending order. Using a directed scan, search the keys for the smallest key and then exchange it with $k_1$. Repeat the search using the keys $k_2$ to $k_n$, and exchange the smallest key in this search with $k_2$. After n - 1 applications of this search and exchange process the keys will be correctly sorted. (See table 8.1 for an example of this method of sorting.)

A typical algorithm for this method is given below.

```
procedure SELECTION (key, n); value n; integer n;
                  integer array key;
begin integer i, j, k, c;
      for i:= 1 step 1 until n - 1 do
      begin k:= i;
            for j:= i + 1 step 1 until n do
            if key [j] < key [k] then k:= j;
            comment the smallest key in the list from i to
            n has been found and is in the kth position. It is
            now interchanged with the ith key if necessary;
            if i ≠ k then begin c:= key [i];
                              key [i]:= key [k];
                              key [k]:= c
                          end
      end
end of procedure SELECTION;
```

This method can be analysed fairly simply:

(a) The number of comparisons can easily be calculated since it is n - 1 on first scan, n - 2 on the second scan etc.

$$\text{Total number of comparisons} = (n - 1) + (n - 2) + \ldots + 2 + 1 = \frac{n}{2}(n - 1). \tag{8.3}$$

This method is of order $n^2$ and thus rather worse than the theoretical minimum order $n\log n$.

(b) The number of comparisons is the same for all initial arrangements of the data and so this method does not benefit from any orderings in the data.

(c) The number of interchanges required is usually one for each search step and therefore $(n - 1)$ for the whole algorithm. On some searches (i. e. if $i = k$) no interchanges are required; calculating the average number of these is left as an exercise for the reader (see example [8. 1]).

(d) As can be seen from the algorithm extra storage is only required for the integer variables i, j, k, c by this method.

| Initial keys | 1st stage | 2nd stage | 3rd stage | 4th stage | 5th stage | 6th stage |
|---|---|---|---|---|---|---|
| 10 | 1 | 1 | 1 | 1 | 1 | 1 |
| 12 | 12 | 4 | 4 | 4 | 4 | 4 |
| 8 | 8 | 8 | 6 | 6 | 6 | 6 |
| 1 | 10 | 10 | 10 | 8 | 8 | 8 |
| 6 | 6 | 6 | 8 | 10 | 10 | 10 |
| 19 | 19 | 19 | 19 | 19 | 19 | 12 |
| 4 | 4 | 12 | 12 | 12 | 12 | 19 |

Table 8. 1. Selection method

**Repeated Selection Sort.** This method is a more sophisticated variation of the simple selection method just described. In its next simplest variation (usually called Quadratic selection) the method is to divide the n keys into about $\sqrt{n}$ groups of about $\sqrt{n}$ keys. The smallest key from each group is selected and stored in a list L. Next, the smallest key in the list L is selected and this is the winner on first round. In the special case where $\sqrt{n}$ is an integer the number of comparisons for this operation is exactly $(\sqrt{n} - 1)$ for each group and there are $\sqrt{n} + 1$ groups (the extra group being list L). Thus the number of comparisons is $(\sqrt{n} + 1)(\sqrt{n} - 1) = n - 1$ which is the same as in ordinary selection. However, in the subsequent steps far fewer comparisons are required because we only need to examine the group from which the smallest key came. The other keys in list L are unchanged. The number of comparisons for this second stage is thus $2(\sqrt{n} - 1)$ as we select in only the group from which the smallest key came and list L. The remaining $n - 2$

steps are similar giving a total number of comparisons = n - 1 + (n - 1) $2(\sqrt{n} - 1) = (n - 1)(2\sqrt{n} - 1)$.

This is an improvement on the simple selection method although a certain amount of extra storage space ($\sqrt{n}$) is required. We can extend these ideas to cubic selection, quartic selection etc. in each case we require fewer comparisons but more storage and a greater complexity of program. The limit of these repeated selection methods is reached when binary comparisons are made at each stage. This method has been called replacement selection but a better name used by Flores is tournament sort. What we have in effect is two keys matched against each other in a similar manner to a knock-out tournament. (The F. A. Cup or Wimbledon Tennis Tournament are examples.) The number of comparisons in tournament sort is of the order n log n so from this point of view it is a very effective method. It has a very close affinity to the methods called treesort and heapsort which also use binary comparisons.

**Exchanging.** There are many methods which sort by exchanging keys; the one that will be described in this section is called Bubblesort because its basic action is to bubble the keys up the list until they meet a smaller key. In particular the smallest key is bubbled to the top on the first pass. In the first pass we start by comparing the nth and (n-1)th keys and exchanging them if necessary and we continue up the list of keys until we compare the second and first items and exchange them if they are in the wrong order. Thus a key is bubbled up until it meets a smaller key, it is then left and the smaller key bubbled up. In the second pass we work from the nth key to the second key, since the first key must, by the first pass, be the smallest key. Although in general (n - 1) passes are necessary to ensure that the list of keys is completely sorted we can make our algorithm more efficient for ordered data by including in each pass an indicator variable (or flag). This flag is set true at the start of a pass and changed to false if any exchange is made. Thus we can terminate the algorithm whenever the flag is true at the end of a pass, since no exchanges have been made and the keys are therefore already ordered. This is illustrated by the example in table 8.2 where seven keys are sorted. In general six passes are necessary but on the fifth pass no exchanges take place and so we can stop. In table 8. 2 there is a staircase of lines, above which no keys are examined on that pass.

| Initial keys | 1st Pass | 2nd Pass | 3rd Pass | 4th Pass | 5th Pass |
|---|---|---|---|---|---|
| 10 | 1 | 1 | 1 | 1 | 1 |
| 12 | 10 | 4 | 4 | 4 | 4 |
| 8 | 12 | 10 | 6 | 6 | 6 |
| 1 | 8 | 12 | 10 | 8 | 8 |
| 6 | 4 | 8 | 12 | 10 | 10 |
| 19 | 6 | 6 | 8 | 12 | 12 |
| 4 | 19 | 19 | 19 | 19 | 19 |

Table 8. 2. Bubblesort method

A typical algorithm for bubblesort is given below:

```
procedure BUBBLESORT (key, n); value n; integer n;
        integer array key;
begin   integer i, j, temp; boolean flag;
        for i:= 1 step 1 until n - 1 do
        begin comment i is the number of passes which is at
                most n - 1, at the beginning of each pass flag is
                set true. The key being bubbled up is kept in temp;
                temp:= key [n]; flag:= true;
                for j:= n step -1 until i + 1 do
                if key [j - 1] < temp then
                    begin comment the key being bubbled up
                            (temp) has encountered a smaller key;
                              key [j]:= temp;
                              temp:= key [j - 1]
                    end
                else begin comment the key is bubbled higher up
                            the list so in effect an exchange takes
                            place and the flag is set false;
                              key [j]:= key [j - 1];
                              flag:= false
                    end;
```

```
                    key [i]:= temp;
                comment if at the end of this pass the
                flag is still true then no exchanges have
                taken place and the keys are in correct
                ascending order;
                if flag then goto EXIT
        end i loop;
EXIT: end of procedure BUBBLESORT;
```

In bubblesort the number of passes will be equal to the maximum number of places between a key in its initial and final positions. In the example in table 8.2 the key 4 moves the most places (5).

The analysis of bubblesort is not as simple as that of selection since the number of passes depends on the data:

(a) If we assume the maximum number of passes is $(n - 1)$ then the number of comparisons is the same as the selection method $= \frac{n}{2}(n - 1)$. For randomly ordered data it can be shown that the expected number passes is less than $(n - 1)$ but since the passes saved are those which contain the fewest number of comparisons the practical effect is not large. A more detailed analysis still gives the most significant term in the expected number of comparisons as $n^2/2$.

(b) The main advantage of bubblesort is that it can be very quick with data which are nearly ordered in the sense that no item is far from its final position.

(c) The number of interchanges required by bubblesort is considerably more than in the previous selection method. For each pass the expected number of interchanges will be approximately half the number of comparisons. For $n - 1$ passes this gives the number of interchanges as

$$n(n - 1)/4$$

which is a slight overestimate of the expected number of interchanges for random data.

(d) Bubblesort requires very little extra storage space.

There are many variations of the basic idea of exchanging keys;

for example we can start at the top of the list and work down, or we can combine these two ideas and perform successive passes of the exchange sort in alternate directions. Variations on this idea called the odd-even transposition sort and Funnel sort are described in examples [8. 14] and [8. 15].

**Insertion.** In this sorting method we assume that when we come to process the ith key the first i - 1 keys are correctly ordered. We now search the i - 1 keys to determine the correct position to place the ith key and move the keys so that we can place the new key in its correct position in the list of i keys. The simplest, but not the most efficient, method of doing this search is to use a directed scan. An algorithm for doing this is:

Compare the new key with the next key in the list (call it the jth key, j starting at i - 1). If the new key is less than jth key move jth key down one place and compare new key to the next key in the list. If the new key is however greater than the jth key we insert it in list at this position. This method is illustrated by the example in table 8. 3, at each pass the → indicates where the new key is inserted in the list. A typical algorithm for this simple insertion sort is given below.

```
procedure INSERTION SORT (key, n); value n;
        integer n; integer array key;
begin integer z, i, j;
        for i:= 2 step 1 until n do
        begin z:= key [i];
                comment z is the current key which has to be
                inserted in its correct position in the list;
                for j:= i - 1 step -1 until 1 do
                if z < key [j] then
                        key [j + 1]:= key [j]
                else begin comment insert the current key z in
                    its correct position in the list of keys;
                    key [j + 1]:= z; goto L
                    end;
```

```
            key [1]:= z; comment this is the case where current
                                 key is top of the list;

      L: end
end of procedure INSERTIONSORT;
```

| Initial keys | 1st pass | 2nd pass | 3rd pass | 4th pass | 5th pass | 6th pass |
|---|---|---|---|---|---|---|
| 10 | 10 | → 8 | → 1 | 1 | 1 | 1 |
| 12 | → 12 | 10 | 8 | → 6 | 6 | → 4 |
| 8 | | 12 | 10 | 8 | 8 | 6 |
| 1 | | | 12 | 10 | 10 | 8 |
| 6 | | | | 12 | 12 | 10 |
| 19 | | | | | → 19 | 12 |
| 4 | | | | | | 19 |

Table 8. 3. Insertion Method

The above algorithm is not difficult to analyse and leads to the expected numbers both of comparisons and of interchanges being of the order $n^2/4$, and clearly little extra space is needed. The number of comparisons will be reduced if the data is in ascending or nearly ascending order.

The number of comparisons can be considerably reduced to about $n \log n$ if instead of using a directed scan to find where to insert the new key we use binary search (see section 7. 2. 2). This produces a sorting method with a number of comparisons of the same order as the minimum possible but with the disadvantage that it requires a large amount of key transfers. So the effect of a high level language compiler could well be to produce bodies of code for these simple but theoretically inefficient methods which have little to choose between them, despite the differences expected from the analyses (or even to give an observed efficiency order quite different from the theoretical). When there are only a few items to sort, a simple method which uses a simple data structure like an array can be the best choice.

## 8.2.2 Binary Tree Sorting Methods

The methods we wish to examine in this section will, like insertion with binary search, be of order $n \log n$ in the number of comparisons; they essentially build a binary sequence search tree (see section 7.2.3). Recently W. A. Martin has called these methods examples of distributive sorts, since in each step the keys are partitioned into two or more subsets. The most widely used of these methods is quicksort, which is described first, followed by other methods closely related to it, each of which sheds a different light on the processes.

**Quicksort.** Given a set of keys $k_1, k_2, k_3, \ldots k_n$ the quicksort method ranks the keys with respect to the first key $k_1$. By interchanges the keys are moved so that all the keys to the left of $k_1$ are smaller than $k_1$ (even though they are not yet fully sorted) and all the keys to the right of $k_1$ are larger. Then recursively we repeat the algorithm for the list both to the left and right of $k_1$. We stop the recursion when all lists consist of a single key. The ranking of the keys with respect to the first key $k_1$ by interchanges is carried out as follows:

Step 1 — Remove first key $(k_1)$ from the list of keys leaving a space in position 1. Set $t := 1$ and $b := n$.

Step 2 — Scan from key $k_b$ back until we find the first key ($k_i$ say) less than $k_1$. Put $k_i$ in the vacant space leaving $k_i$'s old space vacant. Set $b := i$

Step 3 — Scan from key $k_t$ forwards until we find the first key ($k_j$ say) greater than $k_1$. Put $k_j$ in the vacant space leaving $k_j$'s space vacant. Set $t := j$

Step 4 — Repeat steps 2 and 3 until $t = b$. The keys have then been successfully ranked and $k_1$ is put in the vacant position.

Consider the set of keys we have used with previous sorting methods i.e.

| $k_1$ | $k_2$ | $k_3$ | $k_4$ | $k_5$ | $k_6$ | $k_7$ |
|---|---|---|---|---|---|---|
| 10, | 12, | 8, | 1, | 6, | 19, | 4. |

The algorithm is as follows

Remove 10

-, 12, 8, 1, 6, 19, 4

Scan left starting at $k_7$

4, 12, 8, 1, 6, 19, -

Scan right starting at $k_2$

4, -, 8, 1, 6, 19, 12

Scan left starting at $k_6$

4, 6, 8, 1, -, 19, 12

Scan right starting at $k_3$. During this process $t = b$ so 10 is inserted in vacant space.

The keys have now been ranked with respect to the first key 10 and we now rank recursively the two lists of keys 4, 6, 8, 1 and 19, 12 with respect to their first keys.

A detailed algorithm for quicksort is given below using the recursive procedure facility available in Algol.

```
procedure QUICKSORT (key, a, b); value a, b;
        integer a, b; integer array key;
begin integer top, bottom, z;
        if a ≥ b then goto FINISH;
        z:= key [a]; top:= a; bottom:= b + 1;
    UP: bottom:= bottom - 1;
        comment this is the scan from the bottom of the list to
        find a key less than z;
        if bottom = top then goto RECUR;
        if z > key [bottom] then key [top]:= key [bottom]
                                  else goto UP;
  DOWN: top:= top + 1;
        comment this is the scan from the top of the list to find
        a key greater than z;
        if bottom = top then goto RECUR;
        if z < key [top] then begin key [bottom]:= key [top];
                                  goto UP
                             end
                             else goto DOWN;
```

```
RECUR: key [top]:= z;
          QUICKSORT (key, a, bottom - 1);
          QUICKSORT (key, top + 1, b);
FINISH: end QUICKSORT;
```

The local identifiers top and bottom in the procedure start as the ends of the original list but as the scanning process proceeds from both ends their values get closer until finally they are equal. At this point we have found the position for the key being ranked (z) to be inserted.

**Monkey Puzzle Sort.** This method, which is in principle the same as quicksort, works in two stages, the first of which explicitly builds the binary sequence search tree (BSST) for the keys in the same way as in section 7. 2. 3. In stage 2 we traverse the keys of the BSST in symmetric order and thus obtain them in ascending order.

For our particular example the keys

10, 12, 8, 1, 6, 19, 4

give the BSST.

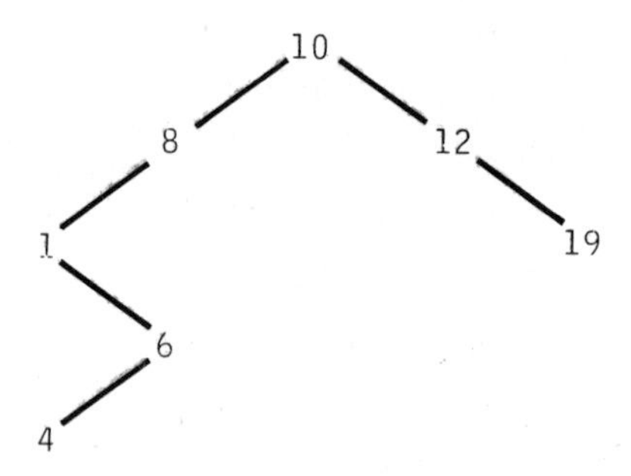


Visiting the above tree in symmetric order we obtain the keys in the order

1, 4, 6, 8, 10, 12, 19.

**Radix Exchange Sort.** If the keys are expressed as binary numbers the relation of the method to quicksort is most easily seen.

Step 1 Scan the most significant bit of the keys starting at the front until a '1' is found. Scan up from the end until the first key with '0' as its most significant digit is found.

Step 2 Exchange the above two keys.

Step 3 Starting from these two keys, repeat the scans up and down on the most significant digit, exchanging two keys when necessary. Stop when the keys have been divided into two sublists that at the front of the original list all starting with the digit '0' and that at the back starting with the digit '1'.

Step 4 Recursively repeat steps 1-3 on the two sublists, scanning on the next most significant digit. Stop when all binary digits have been processed or a sublist has only one key.

Consider the following list of keys:

4, 1, 15, 12, 2, 3, 0, 5, 11.

Table 8. 4 shows columns which express them as binary numbers and the first three stages of scanning to carry out the radix exchange algorithm. The final stage which produces the ordered list is omitted. The divisions between the various sublists are denoted by broken lines, the exchanges by horizontal connected arrows, and the binary digit being considered by a vertical arrow.

| Keys | Binary keys | 1st stage | 2nd stage | 3rd stage |
|---|---|---|---|---|
| 4 | 0100 | 0100← | 0011← | 0000 |
| 1 | 0001 | 0001 | 0011 | 0001 |
| 15 | 1111← | 0101← | 0010 | 0010 |
| 12 | 1100← | 0000 | 0000← | 0011 |
| 2 | 0010 | 0010← | 0101 | 0101← |
| 3 | 0011 | 0011← | 0100 | 0100← |
| 0 | 0000← | 1100← | 1011 | 1011 |
| 5 | 0101← | 1111 | 1111← | 1100 |
| 11 | 1011 | 1011← | 1100← | 1111 |
| | ↑ | ↑ | ↑ | ↑ |

Table 8. 4. Radix Exchange Sort

In this method we can see that if there are n keys and b binary digits in a key the approximate number of exchanges is $\frac{n}{4} \cdot b \doteqdot \frac{n}{4} \log_2 n$ (since $2^b \doteqdot n$ if the key is thick).

Yet another method which builds a binary tree as part of the sorting process is variously known as Treesort and Heapsort. The original algorithms were published by Williams and Floyd but for a good description of the method the reader is referred to a book by Aho, Hopcroft and Ullman. There is also a detailed description of the method in Knuth's book and a certification of Floyd's algorithm has been given by London.

**Analysis of Quicksort or Monkey Puzzle Sort Algorithms.** The analysis of these algorithms is rather longer and less easy than some of the previous ones; it does however introduce a method of deriving some properties of the tree structures arising from random data.

(a) The average number of comparisons. Let $M(n)$ be the average number of comparisons between the keys over all possible orders of the data. Consider the situation for a particular order when the first key $k_1$ has been ranked with respect to the other keys. Suppose that the key $k_1$ is the $(t+1)$th in the final order so that there will be $t$ keys on the left (say) subtree of $k_1$ and $(n - t - 1)$ keys on the right subtree. We will name these two subtrees $S'$ and $S''$ respectively and the whole tree $S$. Most trees $S$ can be produced from more than one order of the data but for any given shape of tree the number of comparisons is the same.

Let $M(n, S)$ be the number of comparisons for $n$ keys leading to a tree of shape $S$. Then every key after the first is compared with $k_1$ ($n - 1$ comparisons) and subsequently any comparisons contribute to those for either subtree $S'$ or $S''$. Hence

$$M(n, S) = n - 1 + M(t, S') + M(n - t - 1, S'') \tag{8.4}$$

Now

$$M(n) = \sum M(n, S)\, P(n, S) \tag{8.5}$$

where $P(n, S)$ is the probability of a tree of shape $S$, and the summation is over all possible shapes of tree. This particular tree $S$ is composed of three independent parts, its root key $k_1$, and the two subtrees $S'$ and $S''$. So $P(n, S)$ is the product of the probabilities of the three components. Probability of $t$ keys on the left subtree = Probability that root is (t+1)th key in final order = $1/n$ since this is the proportion of all the orders which have key $k_1$ presented first.

Thus (8.5) becomes

$$M(n) = \frac{1}{n} \sum_{t=0}^{n-1} \sum\sum M(n, S)\, P(t, S')\, P(n-t-1, S'')$$

where the second and third summations are taken over the different subtrees.

Using the result from (8.4)

$$M(n) = \frac{1}{n} \sum_{t}\sum_{S'} \sum_{S''} [n-1 + M(t, S') + M(n-t-1, S'')]P(t, S')P(n-t-1, S'').$$

We can simplify these terms individually since the sum of the probabilities over all the possible cases is unity, e.g. $\sum_{S'} P(t, S') = 1$. Hence

$$M(n) = \frac{1}{n} \sum_{t=0}^{n-1} (n-1) + \frac{1}{n} \sum_{t}\sum_{S'} M(t, S')P(t, S') + \frac{1}{n} \sum_{t}\sum_{S''} M(n-t-1, S'')P(n-t-1, S'').$$

But the innermost sums of the last two terms are just (8.5) applied to tree of $t$ and $n - t - 1$ items respectively. So

$$M(n) = n - 1 + \frac{1}{n} \sum_{t=0}^{n-1} M(t) + \frac{1}{n} \sum_{t=0}^{n-1} M(n - t - 1),$$

and so, since the last two terms are the same sum counted from opposite directions,

$$M(n) = n - 1 + \frac{2}{n} \sum_{t=0}^{n-1} M(t). \tag{8.6}$$

This equation gives a relation between $M(n)$ and $M(n - 1)$ by subtracting the corresponding equations for $n\,M(n)$ and $(n-1)\,M(n-1)$ and yields the solution

$$M(n) = 2(n + 1) \sum_{i=1}^{n} \frac{1}{i} - 4n \,. \qquad (8.7)$$

If we substitute the approximation $\sum \frac{1}{i} \doteq 0.693 \log_2 n + 0.577$ we obtain

$$M(n) \doteq 1.39(n + 1)\log_2 n - 2.85n + 1.15 \,. \qquad (8.8)$$

As $n$ increases, the largest term in (8.8) is proportional to $n \log_2 n$ and so quicksort is one of the sorting algorithms that gives the order of the number of comparisons for random data equal to the minimum.

(b) Quicksort does not benefit from ordered keys; in fact order in the keys is likely to increase the amount of work since the binary tree forms less branches and becomes very unbalanced. Ideally the binary tree should be close to the complete binary tree and ordered keys hinder such a development. If the keys are in either strictly ascending or descending order then we have the worst cases and the number of comparisons is

$$1 + 2 + 3 + 4 + \ldots + n = n(n + 1)/2 \,.$$

(c) The algorithm for quicksort is implemented so that interchanges as such are avoided. A space is made where the key to be ranked $(k_r)$ was and then keys are shifted until finally the correct position is obtained to reposition $k_r$. The number of key shifts is likely to be about half the number of comparisons or approximately $n \log_e n$.

(d) The extra storage space required by the quicksort algorithm depends on the order in which we process the sublists. When we have completed ranking the first key $k_1$ we then recursively call quicksort to process the two sublists above and below $k_1$. Whichever we decide to do first we must remember the upper and lower bounds of the other sublist. This of course is automatically done for us by the stacking of the actual parameters by the recursive calls. However, it can quite easily be shown by a simple example that the storage requirements are less if at each stage we process the smaller sublist first. (It should be noticed that the detailed algorithm given earlier did not include this refinement.) In effect each new stage of the algorithm gives two lists and if we always sort the smaller first and remember only the division point between the sublists then an induction proof shows that the extra storage can be restricted to at most $\log_2 n$ locations. (See example [8.7].)

**Discussion of Quicksort.** Quicksort is a very effective algorithm for internal sorting particularly if the keys are random. The method has therefore received much attention and extra refinements have been suggested to improve its performance. One suggestion is to use a different method if the sublists are small, for example to use bubblesort for sublists of less than about twelve or even just to compare keys in sublists of two items, interchanging if necessary.

Another idea is to try to remove the disadvantage of an uneven division between the sublists. In other words to choose a key near the median against which to rank the other keys instead of merely taking the first key in each list. A method known as Samplesort has been suggested which estimates the median by sorting a sample, size j, of the n keys. The median of the sample of j keys is taken for the first list and the next sublists use the upper and lower quartile points of the sample. The optimum sizes j of the samples have been calculated assuming that the samples are sorted using quicksort.

### 8.2.3 Sorting by Address Calculation

In address calculation sorting we convert the key by some key transformation into an address in the ordered output list - sometimes called the home address. Therefore if the method is to be effective there must be available a suitable key transformation function. In its simplest form called pigeonhole sorting, we choose a key transformation function which transforms every key into a unique home address in the output list. After that we only have to compress the output list to eliminate holes. The disadvantage clearly is the demand on storage.

In the more general case of sorting by address calculation the key transformation or 'sorting' function will not transform every key into a unique address. The general algorithm is

1. Mark all the locations in the output list (e.g. if all the keys are greater than zero then zero can be used as a suitable indicator of an empty location).

2. Calculate the address(A) of the key(K) to be sorted using the sorting function.

3. If the location A is empty insert the key here and goto step 2.

4. If K > key at location A then search forwards

(step 5)

else search backwards

(step 6)

5. Search forwards. Search until either an empty location is reached in which case insert K here and goto step 2 or a key larger than K is encountered. In this case K is inserted in place of the larger key and this and the subsequent keys are moved forward one by one until an empty location is found. Goto step 2.

6. Search backwards. Search until either an empty location is reached in which case insert K here and goto step 2 or a key is encountered smaller than K. In this case K is inserted in place of the smaller key and this and subsequent smaller keys are moved back one by one until an empty location is found. Goto step 2.

In steps 5 and 6 we can overflow the output list at either end and this will need a test in the algorithm.

This sorting method has obvious similarities to searching using key transformation where collisions are resolved by the linear method (see section 7. 3. 2(a)). The differences are that the hashing method used in searching does not need to put the keys in order in the scatter storage table. Thus the sorting function we use in the address calculation sort is considerably more restricted than the hashing method in searching, but like key transformation in searching, the address calculation method is efficient if the sorting function distributes the keys evenly over the output list.

**Example.** Consider 16 keys supposed approximately evenly distributed in the range 00-99. It is best to allow about twice the space for the keys if possible so as to avoid too much shifting and to avoid congestion at the ends by leaving extra space there. If we assume that 32 locations are available, numbered 1-32, and plan to start with two locations free at each end; an appropriate sorting function is $2 + [\frac{28k_i}{100}]$.

For example, the following random keys

12, 36, 72, 38, 91, 31, 50, 69, 07, 21, 18, 71, 08, 40, 58, 59

will map into the 32 locations shown in table 8. 5 using the algorithm given above.

| Location | Key | Location | Key | Location | Key | Location | Key |
|---|---|---|---|---|---|---|---|
| 1 | | 9 | | 17 | | 25 | |
| 2 | | 10 | | 18 | 58 | 26 | |
| 3 | | 11 | 31 | 19 | 59 | 27 | 91 |
| 4 | 07 | 12 | 36 | 20 | 69 | 28 | |
| 5 | 08 | 13 | 38 | 21 | 71 | 29 | |
| 6 | 12 | 14 | 40 | 22 | 72 | 30 | |
| 7 | 18 | 15 | | 23 | | 31 | |
| 8 | 21 | 16 | 50 | 24 | | 32 | |

Table 8. 5. Address Calculation Sort

The first clash occurs when we try to insert key 71 in location 22, as this location is then occupied by key 72. We therefore search backwards and find location 21 is occupied by key 69 so we insert key 71 here and move 69 down one location. Other collisions occur when we try to insert key 08 in location 4 and key 40 in location 13.

The speed of the address calculation sort depends on

(a) the loading factor (or fullness ratio $\alpha$), and

(b) the effectiveness of the sorting function.

The loading factor, $\alpha = n/N$, where n is the number of keys and N the amount of storage space available for the output list. It was suggested in the example that this ratio should be about 0.5. Address calculation sort is similar in essence to searching using key transformation and linear probing. So if the storage allocated for the output list is nearly equal to the number of keys then the effectiveness of address calculation sort decreases rapidly.

The sorting function is a many to one transformation from the universe of the keys into the output list storage area. Thus an effective sorting function is data dependent which is a disadvantage if we require a generalized sorting routine since there is no general sorting function which is universally efficient. Despite this snag, address calculation is a very important method since it can, with knowledge of the keys, break the $n \log n$ barrier for the number of comparisons. Of course, the calculation of an address is theoretically a more complicated operation than a key comparison but, in practice, it may only be slightly slower. Thus, like key transformation methods in searching, address calculation sorting can be very fast. Another place where address calculation sort can be very useful is in page oriented computers. Here a large penalty is often paid for bringing in a new page and address calculation sorting where adjacent keys are kept on the same page can work very well.

## 8.3 MERGING METHODS

In the previous section internal sorting methods suitable for fast random-access storage were described. In this section the methods described will be applicable to other types of storage, for example magnetic tapes. Some of the methods could also be used quite effectively internally, but certain merging methods are designed to utilise specific characteristics of the tape drives such as sequential access and the ability to read and/or write on the tapes in one or two directions.

### 8.3.1 Balanced Two-Way Merge

The idea of merging lists of keys is a very basic one in sorting. At its simplest, the Balanced Two-way Merge, it consists of progressively combining from the beginning pairs of sorted lists into a longer list, and repeating until just one list remains.

Suppose the unsorted keys are in a list. Divide the list into two lists A and B of as near equal length as possible. Suppose there is storage space available for two other lists C and D.

The method is:

1. Compare the first keys from A and from B and output them in order on C.

2. Compare the next key from A with that from B and output them in order on D.

Repeat steps 1 and 2 until we have strings of length 2 on C and D. Now we merge these strings of length 2 to obtain strings of length 4 on A and B.

3. Compare first key from C and from D and output smallest to A and replace it for comparison purposes by the next key from the string it left.

4. Repeat the comparison in step 3 until one string is exhausted then put the remainder of the other string on A.

Repeat steps 3 and 4 until we have strings of length 4 on A and B. Now we merge these strings to obtain strings of length 8 on C and D and so on until we obtain finally a sorted list of keys.

Example

Starting with keys

18, 23, 02, 50, 42, 63, 20, 28, 33, 03, 47.

Split them into two lists A and B:

A is 18, 23, 02, 50, 42

B is 63, 20, 28, 33, 03, 47

Merge to obtain strings of length 2 on C and D:

C '18, 63,' 02, 28,' 03, 42'

D '20, 23,' 33, 50,' 47

Merge to obtain strings of length 4 on A and B:

A '18, 20, 23, 63', 03, 42, 47'

B '02, 28, 33, 50'

Merge to obtain strings of length 8 on C and D:

C '02, 18, 20, 23, 28, 33, 50, 63,'

D '03, 42, 47'

Merge finally to obtain the sorted string

A 02, 03, 18, 20, 23, 28, 33, 42, 47, 50, 63.

This balanced two way merge is quite effective and can be easily analysed. However, a simple improvement can be made which will take advantage of any ordered strings in the original keys. This is the basis of the natural two-way merging.

### 8.3.2 Natural Two-way Merge

We form two nearly equal lists as before but then

1. Take the first key from A and from B and compare them and put the smaller in C.

2. Replace the key sent to C by the next key from the list (A or B) from which it came.

3. Continue comparing two keys and sending the smaller to C until it is less than the one sent previously.

4. Now send the larger key to C until it also is less than the last one sent to C. When this happens, repeat the operations 1-4 but send the keys to list D.

5. Continue building ordered strings of keys on C and D until A and B are both exhausted.

6. If there are no strings on D then list C contains the ordered list of keys. Otherwise exchange the names of lists C and A, similarly D and B, and go back to step 1.

In the previous example, the first lists A and B are merged onto lists C and D to obtain:

C 18, 23, 63, 03, 42, 47

D 02, 20, 28, 33, 50

Since D is not empty, we repeat the merging process on these two new strings and obtain:

C 02, 18, 20, 23, 28, 33, 50, 63

D 03, 42, 47

D is still not empty so another merging pass is undertaken:

C 02, 03, 18, 20, 23, 28, 33, 42, 47, 50, 63

D empty

The natural two way merge is finished since D is empty and the sorted list of keys is on C.

A detailed algorithm for merge sorting is quite long since so many special cases arise (some ideas which would be useful in writing such an algorithm are given in the hints to example [8. 9]).

**Analysis of Merge Sorting.** The number of passes in balanced two-way merge is fixed; if $2^{k-1} < n \le 2^k$ then there are k passes. Thus the number of passes is $\lceil \log_2 n \rceil$ (i. e. the 'ceiling' of $\log_2 n$, the next greater integer when $\log_2 n$ is not integral and otherwise the value of $\log_2 n$ itself). At each pass we move all n keys so the number of key transfers is $n \lceil \log_2 n \rceil$.

The number of comparisons at each pass is considerably less than n. When the strings are of length 1 we need $\lceil \frac{n}{2} \rceil$ comparisons to merge them into strings of length 2. In the next pass we need at most $\lceil \frac{3n}{4} \rceil$ comparisons and on average $\frac{2n}{3}$ comparisons.

At the next pass the maximum number of comparisons is $\lceil \frac{7n}{8} \rceil$ and the average number $\frac{4n}{5}$; and so on.

Summing over all the $\log_2 n$ passes the maximum number of comparisons is

$$\frac{n}{2} + \frac{3n}{4} + \frac{7n}{8} + \frac{15n}{16} + \dots$$
$$= \sum_{i=1}^{\log_2 n} \frac{(2^i - 1)n}{2^i} = n\left[\frac{n \log_2 n - n + 1}{n}\right] \doteqdot n \log_2 \left(\frac{n}{2}\right) \tag{8. 9}$$

The average number of comparisons, which is difficult to derive, is the sum of (see example [8. 10])

$$\frac{n}{2} + \frac{2}{3}n + \frac{4}{5}n + \frac{8}{9}n + \dots$$

and is very close to the value given in (8. 9).

The natural two-way merge is an improvement over the balanced two-way merge in that it takes advantage of any strings in the original keys. Thus if the original keys are partly sorted having been derived, for example, from sets of keys which have been ordered separately, the natural two-way merge will be much faster than the balanced two-way merge. Even for random keys the natural two-way merge saves about one pass since there are $n/2$ strings on average in a list of $n$ keys. However, the natural two-way merge has the disadvantage that the length of the strings at any pass is not known in advance and it is difficult to reserve the correct space for each string. Thus in general it requires more space than the balanced two-way merge where the length of the strings is known exactly at each pass.

Both the merging methods discussed can be simply extended to become multiway merges. If we have $p$ input lists we can compare $p$ keys and on the first pass we obtain $n/p$ strings of length $p$ which we then distribute on the $p$ output lists. The next pass uses these $p$ output lists as input. Thus we need $2p$ lists and the number of passes is $\log_p n$ for the balanced multiway merge and $\log_p(\frac{n}{2})$ for the natural multiway merge.

Merging methods have been widely used for sorting keys kept on magnetic tapes. The merging methods explained previously are well suited to magnetic tape input and output. However, even more sophisticated merging methods have been developed for use with magnetic tapes and one of the more useful ones will be briefly described.

### 8.3.3 Polyphase Sort

This merging method makes as many tapes available as possible in order to do high order merges. For example, suppose we have four tape-drives and need to sort the 17 keys

12, 36, 72, 38, 91, 31, 50, 69, 07, 21, 18, 71, 08, 40, 58, 59, 97.

The first 7 keys are placed on Tape 1, the next 6 keys on Tape 2 and the remaining 4 on Tape 3:

| | |
|---|---|
| Tape 1 | 12, 36, 72, 38, 91, 31, 50 |
| Tape 2 | 69, 07, 21, 18, 71, 08 |
| Tape 3 | 40, 58, 59, 97 |

Tape 4 is left empty so we merge four strings of length 1 from Tapes 1, 2, 3 to form 4 strings each of length 3 on Tape 4:

| | |
|---|---|
| Tape 1 | 91, 31, 50 |
| Tape 2 | 71, 08 |

| | |
|---|---|
| Tape 3 | empty |
| Tape 4 | 12, 40, 69, ⁞ 07, 36, 58, ⁞ 21, 59, 72, ⁞ 18, 38, 97 ⁞ |

Now merge 2 strings from Tapes 1, 2 (of length 1) and 4 (of length 3) to form two strings (of length 5) on Tape 3:

| | |
|---|---|
| Tape 1 | 50 |
| Tape 2 | empty |
| Tape 3 | 12, 40, 69, 71, 91, ⁞ 07, 08, 31, 36, 58 ⁞ |
| Tape 4 | 21, 59, 72, ⁞ 18, 38, 97 ⁞ |

Now merge 1 string from Tapes 1, 3, 4 to form a string on Tape 2:

| | |
|---|---|
| Tape 1 | empty |
| Tape 2 | 12, 21, 40, 50, 59, 69, 71, 72, 91 |
| Tape 3 | 07, 08, 31, 36, 58 |
| Tape 4 | 18, 38, 97 |

Finally merge 1 string from Tapes 2, 3 and 4 to form the completely sorted string on Tape 1:

| | |
|---|---|
| Tape 1 | 07, 08, 12, 18, 21, 31, 36, 38, 40, 50, 58, 59, 69, 71, 72, 91, 97 |

| Tapes | | | Total number of strings |
|---|---|---|---|
| 0 | 0 | 1 | 1 |
| 1 | 1 | 1 | 3 |
| 1 | 2 | 2 | 5 |
| 2 | 3 | 4 | 9 |
| 4 | 6 | 7 | 17 |
| 7 | 11 | 13 | 31 |
| 13 | 20 | 24 | 57 |
| ⋮ | ⋮ | ⋮ | ⋮ |

Table 8.6. Polyphase sort with 4 tapes. The perfect number of strings at each stage.

The initial distribution of keys was chosen so that on the penultimate step we had one sorted string on three of the four tapes, and at each pass the maximum number of strings were available for a three-tape merge. Working from the final step we can calculate the number of strings required at each stage for a polyphase merge. For example, with four tapes we will always merge using three of the tapes and the figures are given in table 8.6.

In general if we have t tapes the distribution of the strings on the (t - 1) tapes is determined by the following three rules

1. First row is all zeros except for a 1 in the last column

$$U_{1,j} = 0, \quad j = 1, 2, \dots t\text{-}2, \; U_{1,t-1} = 1.$$

2. The first number in the ith row is equal to the last number in the (i-1)th row:

$$U_{i,1} = U_{i-1,t-1}.$$

3. The jth number in the ith row is the sum of the (j-1)th number in (i-1)th row and the first number in the ith row

$$U_{i,j} = U_{i-1,\; j-1} + U_{i,1}$$

for $j = 2, 3, \dots t-1$.

The total number of strings at any stage is the sum of the (n - 1) previous numbers. (For example, for $n = 4$ tapes $57 = 31 + 17 + 9$.) In general the number of strings to be merged will not fit exactly into the required pattern. In this case we must add dummy strings at the input stage to achieve the required numbers in a row of the table. These dummy strings need not exist on the tape and their processing can be avoided. As a refinement the best positioning of these dummy strings can be found to minimize the amount of processing. Clearly an improvement in the method as described can be achieved by working initially with the number of strings present in the data.

The polyphase sorting method as described above requires the tapes to be rewound. If backward reading is permitted with the magnetic tapes then the algorithm can be revised so that the tapes are not rewound at each pass.

Polyphase sort is regarded by many as the best known method if the number of magnetic tapes is six or fewer. For more than six tapes the oscillating merge sort is superior, but as such large tape sorting operations are now confined to rather special applications we merely give an appropriate reference.

## 8.4 A COMPARISON OF SORTING METHODS

There are a great number and variety of sorting methods and only a few of these have been examined in this chapter. Unfortunately there is such a diversity of different applications of sorting methods that we can neither point out one method which is universally effective nor eliminate methods which are universally bad. Simple methods tend to need a larger number of operations ($kn^2$ as against kn log n) compared with the more sophisticated methods, but the operations themselves and the house-keeping of the program are simpler and for moderate numbers of keys a simple method may outperform those theoretically more efficient. Of course, just what constitutes a 'moderate' number of keys in this context depends upon the computer configuration, the compiler and the control program operating.

One such comparison - sorting 100 random keys using various algorithms in ALGOL W and its compiler on an IBM 360/67 operating under the Michigan Timesharing System control program - is given in table 8.7 and shows that generally quicksort is a very fast method for random keys. Its main drawback is not only that it takes no advantage of ordered keys but it is in fact slower in these cases. It is interesting to note that the simple selection method and insertion with binary search gave the same time. In fact, with the ALGOL W compiler used, the extra time that selection took to do the comparisons was balanced by the smaller time it took to move the keys. Bubblesort, although slow in general, would improve if the data was ordered. For ordered or semi-ordered data probably the fastest method would be natural two-way merge, which does however require extra storage space whilst bubblesort is practically done in situ.

Table 8. 7. Comparison of sorting methods used internally

| Sorting method | Key comparisons | | | Key transfers or interchanges | | | Extra storage | Effect on method of ordered data | Time (ALGOL W - 100 keys) |
|---|---|---|---|---|---|---|---|---|---|
| | Average | Max. | Min. | Average | Max. | Min. | | | |
| Selection | $\frac{n^2}{2} - \frac{n}{2}$ | $\frac{n^2}{2} - \frac{n}{2}$ | $\frac{n^2}{2} - \frac{n}{2}$ | $\sim(n-1)$ | n-1 | 0 | Negligible | Slightly fewer key transfers | 0. 21 |
| Quadratic selection | $(n-1)(2\sqrt{n}-1)$ | $(n-1)(2\sqrt{n}-1)$ | $(n-1)(2\sqrt{n}-1)$ | $\sim(n-1)$ | n-1 | 0 | $\sim\sqrt{n}$ | Slightly fewer key transfers | - |
| Bubblesort | $\sim(\frac{n^2}{2} - \frac{n}{2})$ | $(\frac{n^2}{2} + \frac{n}{2})$ | n | $\sim\frac{n(n-1)}{4}$ | $\frac{n(n-1)}{2}$ | 0 | Negligible | Improved | 0. 40 |
| Insertion | $\sim\frac{n^2}{4}$ | $\sim\frac{n^2}{2}$ | n | $\frac{n(n-1)}{4}$ | $\frac{n(n-1)}{2}$ | 0 | Negligible | None | 0. 25 |
| Insertion with binary search | $\sim n \log_2 n$ | $\sim n \log_2 n$ | n | $\frac{n(n-1)}{4}$ | $\frac{n(n-1)}{2}$ | 0 | Negligible | None | 0. 21 |
| Quicksort (treesort) | $\sim 1.39 n \log_2 n$ | $\sim\frac{n^2}{2}$ | $\sim 0.5 \log_2 n$ | $\sim 0.7 n \log_2 n$ | - | 0 | $\sim \log_2 n$ | Worse for ordered data | 0. 09 (0. 13) |
| Address calculation | $\frac{n(1-\alpha/2)}{(1-\alpha)}$ | $\sim\frac{n^2}{4}$ | n | $n+\frac{n(1-\alpha/2)}{(1-\alpha)}$ | $\sim\frac{n^2}{4}$ | n | $\sim\frac{n}{4}$ | None depends on key distribution | - |
| Natural two-way merge | $\sim n \log_2 (\frac{n}{2})$ | $\sim n \log_2 n$ | n | $\sim n \log_2 (\frac{n}{2})$ | $\sim n \log_2 n$ | n | 2n | Improved | 0. 14 |

n is the number of keys; $\alpha = \frac{n}{N}$ where N is the size of output list (or storage table) in address calculation sort.

~ indicates an approximate value

## 8.5 BIBLIOGRAPHY

There is a vast literature on sorting and only some of the more important books and articles will be referenced here. Since the first edition of this book was published the third volume of Knuth's The Art of Computer Programming has appeared [1]. The section on Sorting is both detailed and comprehensive and Knuth's book is very useful as a general reference to sorting methods. Another recent book on sorting by H. Lorin [2] describes many sorting methods and contains most of the detailed CACM sorting algorithms as an appendix. The general survey of sorting methods by W. A. Martin [3] is still very readable and useful. The older books on sorting by I. Flores [4] and K. E. Iverson [5] have considerable notational difficulties. Flores writes his algorithms in the assembly language FLAP and Iverson's algorithms are described in the original APL; however Iverson is one of the few authors, apart from Knuth, who attempts to analyse mathematically the sorting algorithms. Although we have classified sorting methods into 'selection', 'exchange', 'insertion', 'binary tree', 'address calculation', and 'merging', such distinctions are not always clear cut nor universally recognised. For example Quicksort considered here as a binary tree method is also a partition exchange method, and Heapsort can be considered either as tournament selection or a binary tree method.

Several other references are of interest for particular sorting methods. Quadratic selection was first published by Friend [6] who showed how it could be extended to cubic, quartic etc. The limit, which Friend called '$n^{th}$ degree selection' and which Knuth calls 'tree selection', is also known as tournament sort. Heapsort was discovered by Williams [7] and a very lucid description of this method is to be found in Aho, Hopcroft and Ullman [8]. The related method of Treesort was suggested by Floyd [9] and a detailed certification of this method by London [10] gives a good description of it.

Windley [11] introduced the idea of monkey puzzle sort and Hoare [12] that of quicksort, and Hibbard [13] has some interesting algorithms for these methods. Hildebrandt and Isbitz [14] have discussed in detail radix exchange sort. The extensions to quicksort are numerous and those of particular interest are Frazer and McKellar [15], who introduce the ideas of samplesort, van Emden [16], and the algorithms of Scowan [17] and Singleton [18].

Sorting by address calculation was studied by Isaac and Singleton [19] and an algorithm called Mathsort has been produced by Feurzeig [20].

Merge sorting is one of the oldest methods and goes back at least as far as John von Neumann in 1945. Polyphase merge was first discovered by Gilstad [21] and oscillating merge by Sobel [22]. Finally a useful method not discussed in this chapter is shellsort which is named after its discoverer [23]; there is an algorithm for this method by Boothroyd [24].


[1] D. E. Knuth: The Art of Computer Programming, Volume 3: Sorting and Searching, pp. 1-388, Addison-Wesley, 1973.

[2] H. Lorin: Sorting and Sort Systems. Addison-Wesley, 1975.

[3] W. A. Martin: Sorting, ACM Computing Surveys, vol. 3, no. 4 (December 1971), pp. 147-74.

[4] I. Flores: Computer Sorting, Prentice-Hall Inc., 1969.

[5] K. E. Iverson: A Programming Language, chapter 6, pp. 176-245, John Wiley and Sons, 1962.

[6] E. H. Friend: Sorting on Electronic Computer Systems, J. A. C. M. Vol. 3, No. 3, (July 1956), pp. 134-68.

[7] J. W. J. Williams: A. C. M. Algorithm 232: Heapsort, Comm. A. C. M., Vol. 7, No. 6, (June 1964), pp. 347-8.

[8] A. V. Aho, J. E. Hopcroft and J. D. Ullman: The Design and Analysis of Computer Algorithms, Addison-Wesley, 1974, pp. 87-92.

[9] R. W. Floyd: A. C. M. Algorithm 245: Treesort 3, Comm. A. C. M., Vol. 7, No. 12 (December 1964), p. 701.

[10] R. L. London: Certification of Algorithm 245. Treesort 3: Proof of Algorithms - a new kind of Certification, Comm. A. C. M., Vol. 13, No. 6 (June 1970), pp. 371-3.

[11] P. F. Windley: Trees, Forests and Rearranging, Computer J., Vol. 3, No. 2 (July 1960), pp. 84-8.

[12] C. A. R. Hoare: Quicksort, Computer J., Vol. 5, No. 1 (1962), pp. 10-15.

[13] T. N. Hibbard: Some combinatorial properties of certain trees with application to searching and sorting, J. A. C. M., Vol. 9, No. 1 (January 1962), pp. 13-28; also, An empirical study of minimum storage sorting, Comm. A. C. M., Vol. 6, No. 3 (May 1963), pp. 206-13.

[14] P. Hildebrandt and H. Isbitz: Radix Exchange - an internal sorting method for digital computers, J. A. C. M., Vol. 6, No. 2 (April 1959), pp. 156-63.

[15] W. D. Frazer and A. C. McKellar: Samplesort: A sampling approach to minimal storage tree sorting, J. A. C. M., Vol. 17, No. 3 (July 1970), pp. 496-507.

[16] M. H. van Emden: Increasing the efficiency of Quicksort, Comm. A. C. M., Vol. 13, No. 9 (Sept. 1970), pp. 563-7; also ACM Algorithm 402, Comm. A. C. M., Vol. 13, No. 11 (November 1970), p. 693.

[17] R. S. Scowan: A. C. M. Algorithm 271: Quickersort, Comm. A. C. M., Vol. 8, No. 11 (November 1965), pp. 669-70.

[18] R. C. Singleton: A. C. M. Algorithm 347, Comm. A. C. M., Vol. 12, No. 3 (March 1969), pp. 185-7.

[19] E. J. Isaac and R. C. Singleton: Sorting by Address Calculation, J. A. C. M., Vol. 3, No. 3, (July 1956), pp. 169-74.

[20] W. Feurzeig: A. C. M. Algorithm 23: Mathsort, Comm. A. C. M., Vol. 3, No. 11 (November 1960), p. 601.

[21] R. L. Gilstad: Polyphase Merge Sorting - an advanced technique, Proc. EJCC, Vol. 18, (December 1960), pp. 143-8. Spartan Books, New York.

[22] S. Sobel: Oscillating Sort - a new sort merging technique, J. A. C. M., Vol. 9, No. 3 (July 1962), pp. 372-4.

[23] D. L. Shell: A high speed sorting procedure, Comm. A. C. M., Vol. 2, No. 7 (July 1959), pp. 30-2.

[24] J. Boothroyd: A. C. M. Algorithm 201: Shellsort, Comm. A. C. M., Vol. 8, No. 6 (August 1963), p. 445.

## EXAMPLES 8

[8. 1] Estimate the number of key interchanges that the selection method will require for random keys.

[8. 2] Write a detailed algorithm for the quadratic selection method of sorting.

[8. 3] Estimate the average number of key comparisons and interchanges to sort random data using the insertion algorithm given in section 8. 2. 1.

[8. 4] Write a detailed algorithm for sorting using insertion with binary search.

[8. 5] Given the eight keys

6, 5, 7, 1, 3, 2, 4, 8

how many actual comparisons and key interchanges are required by the following methods?:

(a) Bubblesort,

(b) Insertion,

(c) Insertion with binary search.

Modify the algorithms to obtain counts of the numbers of comparisons and insertions, and produce these counts for larger sets of keys.

[8. 6] Give a detailed algorithm for the natural two-way merge sort.

[8. 7] Show that the average number of comparisons required to merge two ordered strings of $l$ and $m$ items selected at random from the set $l + m$ is $A(l, m) = lm(l + m + 2)/(l + 1)(m + 1)$, by considering the situation after the first comparison has been made and so deriving a relationship between $A(l, m)$, $A(l - 1, m)$ and $A(l, m - 1)$. Show that the probability that exactly $r$ items of the $l$ string are greater than all the items of the $m$ string is $\frac{l!\,m(l+m-r-1)!}{(l+m)!\,(l-r)!}$.

[8. 8] Write down the table for the exact number of strings required at each stage of a polyphase merge if six tapes are available (give eight stages in the table).

[8.9] Suggest a suitable sorting function for use with address calculation sort when the number of keys is 100 and the 200 locations 1-200 are available for the output list. The keys are alphabetic and will be the surnames of people. Demonstrate how your sorting function operates for the following 100 surnames.

Aldridge
Armstrong
Atherton
Atkinson
Aubrey
Badger
Baker
Barker
Beard
Bell
Bennett
Benson
Birthwhistle
Blake
Blakey
Brand
Bray
Brewis
Brown
Bullen
Burgess
Butler
Cabot
Carr
Champion
Close
Collins
Cooper
Davies
Dawson
Dickinson
Eden
Elliott
Emmerson
Fell
Fleming
Fletcher
Ford
Foster
Fraser
Gardner
Gibson
Goddard
Grant
Halliday
Harper
Harvey
Hayes
Henderson
Hill
Houston
Hunter
Inch
Jackson
Johnson
Johnston
Jones
Kelly
King
Laurence
Lewis
Long
Lowther
McLaren
McNaughton
Mansfield
Metcalfe
Mills
Moffat
Mullens
Newton
Parker
Paterson
Patterson
Pearson
Quinn
Reynolds
Richardson
Robinson
Rose
Savage
Scott
Shields
Smith
Stead
Stevenson
Taylor
Thomas
Thompson
Todd
Veitch
Wallace
Ward
West
White
Williams
Wilson
Wood
Wright
Young

[8.10] Records of information including their names and faculties of university students are of fixed length of 64 words; about 250 such records can be held in the computer main store available and leave sufficient space for a processing program. There are four magnetic tape drives attached to the computer system. Alphabetical lists of all students are required together with similar lists for each of seven faculties.

An internal sort is to be used to generate ordered strings for output to the tapes in preparation for subsequent merging.

Say how you would generate the lists required if at most 8000 student records have to be processed. Comment upon the allocation of the strings to the different tape drives.

Describe in detail (giving a flow diagram and supporting commentary) a suitable method of internal sorting.

(Newcastle, M. Sc. 1967)

[8.11] Describe the advantages and disadvantages of the following methods for sorting numbers in a computer memory:

(i) sift sort,

(ii) merge sort (or combined sift-merge sort),

(iii) pigeon-hole or address sorting.

What factors affect the speed of a magnetic tape sort? Explain how they affect it.

Describe briefly the three phases normally present in a program to sort data on magnetic tapes.

(Glasgow, 1968)

[8.12] Draw a flowchart for an algorithm to sort a list of N records into ascending key order. Convert the flow chart into a computer program which accepts a list of N ordered pairs of real numbers as input and produces an order list as output if the first number in each pair is interpreted as the key.

Discuss briefly the algorithm used in relation to the following criteria:

(i) Variation of execution time with the length of the list,

(ii) The data storage space required by the algorithm, and

(iii) The influence on the efficiency of the algorithm of characteristics of the input list.

(Belfast, 1969)

[8.13] Write down a set of recurrence relations which represent the number of strings on each tape at the completion of any pass of a polyphase sort.

The sequence of integers

$$S_1, S_2, S_3, \ldots, S_n, S_{n+1}$$

represents the string distribution after the initial adjustment phase, in readiness for a polyphase sort of order n. Prove, for the n integers which are non-zero, that either

(a) they are all odd; or

(b) one is odd, and the rest are even.

Prove also, that in case (a) the final sorted output will be the empty tape (with 0 strings) and in case (b) the final sorted output tape will be the tape containing an odd number of strings.

(Essex, 1971)

[8.14] The Odd-even transposition sort is described as follows. Each pass consists of two half-stages, in the first half-stage compare each odd indexed key (except the last if it is of odd index) with its successor and exchange them if necessary, in the second half-stage compare each odd indexed key (except the first) with its predecessor and exchange them if necessary. An example is given below.

| Index | Initial keys | 1st pass | | 2nd pass | | 3rd pass | | 4th pass | |
|---|---|---|---|---|---|---|---|---|---|
| 1 | 10 | 10 | 10 | 1 | 1 | 1 | 1 | 1 | 1 |
| 2 | 12 | 12 | 1 | 10 | 6 | 6 | 4 | 4 | 4 |
| 3 | 8 | 1 | 12 | 6 | 10 | 4 | 6 | 6 | 6 |
| 4 | 1 | 8 | 6 | 12 | 4 | 10 | 8 | 8 | 8 |
| 5 | 6 | 6 | 8 | 4 | 12 | 8 | 10 | 10 | 10 |
| 6 | 19 | 19 | 4 | 8 | 8 | 12 | 12 | 12 | 12 |
| 7 | 4 | 4 | 19 | 19 | 19 | 19 | 19 | 19 | 19 |

Give an algorithm or detailed flow diagram for this sorting method and discuss the stopping conditions you use.

Given the ten keys

6, 5, 9, 7, 10, 1, 3, 2, 4, 8

how many actual key comparisons and key interchanges are required by the odd-even transposition method to sort them into ascending order? How do these results compare with those obtained using bubble sort on the above keys?

(Newcastle 1973)

[8.15] Funnel Sort is a sorting method very similar to Bubble Sort except that successive passes are made in opposite directions. On the first pass we start at the bottom of the list comparing the bottom two keys and exchanging them if necessary and work up the list until we compare the second and first keys. On the second pass we start with the second and third keys comparing and exchanging them if necessary and work down the list. Successive passes therefore are made in opposite directions and each pass is one key shorter than the previous one. An example is given below

| Index | Initial keys | 1st Pass | 2nd Pass | 3rd Pass | 4th Pass | 5th Pass | 6th Pass |
|---|---|---|---|---|---|---|---|
| 1 | 5 | 1 | 1 | 1 | 1 | 1 | 1 |
| 2 | 2 | 5 | 2 | 2 | 2 | 2 | 2 |
| 3 | 12 | 2 | 5 | 3 | 3 | 3 | 3 |
| 4 | 9 | 12 | 9 | 5 | 5 | 4 | 4 |
| 5 | 3 | 9 | 3 | 9 | 4 | 5 | 5 |
| 6 | 4 | 3 | 4 | 4 | 6 | 6 | 6 |
| 7 | 1 | 4 | 6 | 6 | 8 | 8 | 8 |
| 8 | 6 | 6 | 8 | 8 | 9 | 9 | 9 |
| 9 | 8 | 8 | 12 | 12 | 12 | 12 | 12 |

Give an algorithm or detailed flow diagram for this sorting method and discuss the stopping conditions you use.

Given the ten keys

5, 10, 2, 9, 1, 3, 6, 8, 7, 4

how many actual key comparisons and key interchanges are required by the Funnel Sort method? How do these results compare with those obtained using Bubble Sort on the above keys?

Discuss which general sets of keys would be more effectively sorted by Funnel Sort than Bubble Sort and vice-versa.

(Newcastle 1974)

[8.16] The method of sorting by counting is based on the fact that if a key is in the $j^{th}$ position in the final list then it is greater than exactly $(j - 1)$ other keys in the list. (Assuming the keys are being sorted into ascending order.) If there are N keys $k_1, k_2, \ldots, k_N$, then the obvious way to do this is to compare $k_i$ with $k_j$ for i and j from 1, 2, ... N; but over half these comparisons are redundant, since we do not need to compare a key with itself nor do we need to compare both $k_i$ with $k_j$ and $k_j$ with $k_i$. Write a detailed algorithm for this method which is reasonably efficient with regard to key comparisons. How many key comparisons would you expect your algorithm to take for N random keys? How is it affected by ordered data?

Given the following set of keys

5, 7, 4, 12, 9, 1, 8, 6, 14

how many key comparisons does your method require? Compare this to the result you would obtain if you used the bubble sort method on this set of keys. How many key interchanges would the bubble sort method require for these keys?

(Newcastle 1975)

[8.17] Describe in detail the method of internal sorting known as Quicksort. Show how your method would sort the eleven keys

6, 4, 9, 10, 1, 11, 3, 5, 2, 8, 7

into ascending order and state how many actual key comparisons and key interchanges are required in this example.

If the number of keys to be sorted $n = 2^i - 1$ and if during the Quicksort algorithm the smaller of the two sublists is always processed first, then prove by induction (or otherwise) that the extra storage space required is about $\log_2 n$ locations.

(Newcastle 1975)

[8.18] Group Sorting Project

This project is a mixture of individual program design and group cooperative effort. Each member of the group will work individually on a different sorting method for which he will design a program and the group will cooperate on the design of test data for comparing the methods. The purpose of the project is to ascertain the overall effectiveness of each method with respect to the others.

Someone marking the projects will be looking for evidence of good program design, understanding of the method and appropriate choice of data structures, intelligent evaluation and a readable report.

In the group project, each student is expected to write a program for one particular sorting method and to get it working satisfactorily. The group as a whole should decide on the methods of overall comparison. They would be expected to test the methods with random keys and partially ordered keys and data sets of sufficient size to give realistic results. They should obtain practical counts of key comparisons and key transfers. In the write-up, students should discuss how their methods performed on (a) random data and (b) partially ordered data and also to compare the efficiency of their method with respect to:

(i) number of key comparisons

(ii) number of record or key transfers

(iii) extra storage space required.

General References for the following sorting methods are given in section 8.5, Bibliography.

<u>Selection Methods</u>

1. <u>Quadratic Selection</u>

   A modification of the simple linear selection. (See section 8.2.1 and also Knuth [1] p. 141, Lorin [2] p. 109 and Flores [4] p. 55.)

2. <u>Cubic Selection</u>

   The next order after Quadratic Selection. (See Flores [4] p. 66.)

3. <u>Tournament Sort</u>

   This is the logical conclusion of higher order selection sorts when two keys only are compared. (See Knuth [1] p. 142, Lorin [2] p. 59, and Flores [4] p. 70.)

4. <u>Heapsort</u>

   This is a tree selection method of sorting and is explained in detail in Aho, Hopcroft and Ullman [8] pp. 87-92 and also in Knuth [1] p. 143.

5. <u>Replacement Selection</u>

   This method, which is also called Replacement Tournament Sort, is described in Friend [6], Martin [3] p. 159 and Flores [4] pp. 113-29.

6. Natural Selection

A method suggested for improving Replacement Selection by getting longer strings of keys before merging. (See W. D. Fraser and C. K. Wong, C. A. C. M., Vol. 15 (Oct. 1972), p. 910, also IBM Report - Report RC 3564.)

Merging Methods

7. Balanced Two-Way Merge

In this merging method the two strings are the same size. (See section 8. 3. 1, Knuth [1] p. 163 and Flores [4] p. 98.)

8. Natural Two-Way Merge

Similar to Balanced Two-Way Merge but the strings are no longer of fixed length. (See section 8. 3. 2, Knuth [1] p. 161, Lorin [2] p. 126, and Flores [4] p. 98.)

Binary Tree Type Methods

9. Modified Quicksort

The basic Quicksort algorithm is well known. It is suggested that it is modified so that if the number of keys to be sorted at any stage $< X$ use a bubble sort. Find the best value of X, it should be about 12. (See Martin [3] p. 151.)

10. Radix Exchange Sort

This method is explained in section 8. 2. 2. See also Knuth [1] p. 125, Martin [3] p. 152, and Hildebrandt and Isbitz [14].

11. Samplesort

Another modification of the Quicksort algorithm. This one ensures that the key to be ranked is well chosen. (See Martin [3] p. 153, Frazer and McKellar [15] and J. G. Peters and P. S. Kritzinger, BIT Vol. 15 (1975) p. 85.)

Other Methods

12. Insertion with Binary Search

An improvement on the simple insertion algorithm in that it finds the correct place to insert the key by a binary search. (See Knuth [1] p. 83, Iverson [5] p. 221 and Flores [4] p. 40.)

13. Address Calculation Sort

In this method, the key is transformed by a sorting function to give an address in storage and is explained in section 8. 2. 3.

(See also Knuth [1] p. 99, Lorin [2] p. 162, Martin [3] p. 156, Flores [4] p. 85 and Isaac and Singleton [19].)

14. Shell Sort

This method is a mixture of merging and exchanging. The basic idea is to combine the small number of comparisons in merging with the economy of storage of exchanging. (See Knuth [1] p. 84, Lorin [2] p. 37, Martin [3] p. 149, Shell [23] and Boothroyd [24].)

# Notes on the solutions to examples

## EXAMPLES 2

[2.1] Two superposed decimal digits can certainly produce another combination representing a digit but they need not do so.

If no holes overlap in the patterns, the parity of the resulting combination will be even and so cannot represent a digit e.g. 1 and 4 lead to 00101.

If a character with three holes is superposed on a character represented by one of those holes the latter appears never to have been punched and the resulting combination is just the three-hole character, e.g. 2 and 3 lead to 10011, i.e. 3. For an erase to be produced by overpunching two digits, we need two three-hole characters which overlap by just one hole, e.g. 7 and 9.

This type of equipment failure is a little awkward to detect from the tape alone (although it would probably be observed by a person using a keyboard perforator) and needs some information about the set of characterms being punched - for example, by a check on the sum of the characters punched, or on the number of digits.

[2.2] (a) Even. (b) Odd. $H(a, b) = 3$.

The Hamming distance between two strings of different parity is always an odd number. One method of finding the distance is to add the binary strings without any carry digits. Accordingly the number of 'ones' in the sum string is (total number of 'ones' in the two strings) - 2(number of places in which 'ones' occur in both strings) and hence is an odd number.

[2.3] (a) $H(w_1, w_2) = H(w_3, w_4) = 4$,

$H(w_1, w_3) = H(w_1, w_4) = H(w_2, w_3) = H(w_2, w_4) = 3$.

(b) $C(w_1) = (11100, 00100, 01000, 01110, 01101)$

$C(w_2) = (10011, 01011, 00111, 00001, 00010)$

$C(w_3) = (00000, 11000, 10100, 10010, 10001)$

$C(w_4) = (01111, 10111, 11011, 11101, 11110)$

No common members

(c) $H(w_1, s) = H(w_2, s) = 3$, $H(w_3, s) = H(w_4, s) = 2$.

(d) The code corrects single errors and detects double errors, e. g. s cannot be decoded uniquely.

[2. 4] No, since the code digits of $W_4$ and $W_5$ are not the same up to the last digit; we can replace these codes by $W_4 \to 110$, $W_5 \to 111$ and shorten the average length of the code (unless $P(W_4) = P(W_5) = 0$, when a similar replacement for $W_2$ and $W_3$ can be made).

[2. 5] $a \to 1$, $b \to 01$, $c \to 0011$, $d \to 0010$, $e \to 00011$, $f \to 00010$, $g \to 00001$, $h \to 00000$.

Other Huffman codes arise from different assignments of 0s and 1s leading to code words of the same length.

$H(S) = L = 2\frac{1}{8}$.

The shortest block code has length 3.

[2. 6] This is one of the examples given in Huffman's original paper. One code has words

10, 000, 011, 110, 111, 0101, 00100, 00101, 01000, 01001, 00110, 001110, 001111.

$L = 3.42$.

[2. 7] (b) and (c).

[2. 8] (b), (d) and (e). There are $2^5 = 32$ possible 5-bit patterns and only one eighth of these are used. $H(s_2, s_3) = 3$.

[2. 9] Three bits (decode on two out of three correct).

Five bits: $m = 2$, $r = 3$ in §2. 4.

[2. 10] Four: use e. g. code of example [2. 3].

[2.11] (a) $2^6 = 64$. (b) $1 + 2(2^6 - 1) = 127$.

All-ones is commonly 'erase' (in both case shifts).

[2.12] Single errors can be detected but <u>not</u> corrected.

[2.13] A → 11, B → 01, C → 101, D → 001, E → 1000, F → 1001, G → 0000, H → 0001.

L = 2.75.

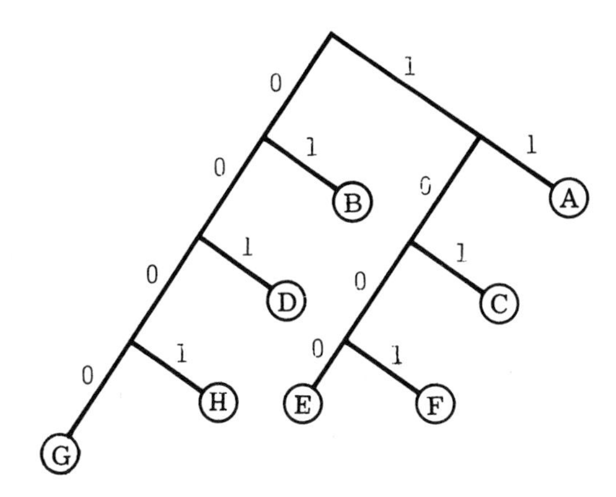


[2.14] There are eight symbols, so that a block code of 3 bits (up to eight combinations) is the shortest possible.

The Huffman code with the longest codes as short as possible is a block code of 3 bits.

In order to keep the shortest possible codes for symbols with the highest probability, we keep these symbols uncombined in the algorithm as long as possible. One such code is

$V_1 \to 01$, $V_2 \to 00$, $V_3 \to 100$, $C_1 \to 110$, $C_2 = 1011$, $C_3 = 1010$, $C_4 = 1111$
$C_5 = 1110$.
L = 3.

[2.15] $P\{r \text{ heads}\} = \binom{10}{r} \cdot \frac{1}{2^{10}}$

Hence P(r=0)=P(r=10)=1/1024, P(r=1)=P(r=9)=10/1024, P(r=2)=P(r=8)=45/1024, P(r=3)=P(r=7)=120/1024, P(r=4)=P(r=6)= 210/1024, P(r=5)=252/1024.

One coding is:

0 → 1111110, 1 → 111110, 2 → 1110, 3 → 011, 4 → 10, 5 → 00, 6 → 010, 7 → 110, 8 → 11110, 9 → 1111110, 10 → 1111111.

$L = 2.76$.

$H(S) = 2.7$.

[2.16] 64 symbols. $H(S) = \sum_{1}^{64} -\frac{1}{64}\log 64 = 6$.

Binary representation of the octal digits e.g. 00 → 000 000, 01 → 000 001, ... 77 → 111 111.

(a) The source with two different octal digits has entropy

$$= \sum_{1}^{56} -\frac{1}{56}\log\frac{1}{56}$$

$$= \log 56$$

Therefore information given $= 6 - \log 56 = 0.2$ bits.

(b) Similarly, information given $= 6 - \sum_{1}^{8}\frac{1}{8}\log\frac{1}{8} = 3$ bits.

(c) $6 + \sum_{1}^{32}\frac{1}{32}\log\frac{1}{32} = 1$ since $40_{(8)} = 32$.

(d) 1.

[2.17] Second extension:

| Symbol | Prob. | Code |
|---|---|---|
| $S_1S_1$ | 9/16 | 0 |
| $S_1S_2$ | 3/16 | 10 |
| $S_2S_1$ | 3/16 | 110 |
| $S_2S_2$ | 1/16 | 111 |

$L = 27/16$ bits per symbol of the second extension $= \frac{27}{32}$ bits/symbol $= 0.84$.

$H(S) = 0.81$.

[2.18] Sequences AAA, AAB, ABA, BAA, ABB, BAB, BBA, BBB, each with prob $= \frac{1}{8}$.

$H(S_1) = 3$.

$S_2$ has four symbols, probs. $\frac{1}{8}, \frac{3}{8}, \frac{3}{8}, \frac{1}{8}$.

$H(S_2) = 1.8$.

[2.19] Assign codes with one mark to the symbols with the three greatest probabilities, and so on.

$S_1 \rightarrow 100$, $S_2 \rightarrow 010$, $S_3 \rightarrow 001$, $S_4 \rightarrow 110$, $S_5 \rightarrow 101$, $S_6 \rightarrow 011$, $S_7 \rightarrow 111$

Average number of marks $= (0.3+0.2+0.15) + 2(0.15+0.15+0.03)+3(0.02)$
$= 1.37.$

[2.20] (a) Form the tens digits of the sum by adding 0011 to the carry digit (i.e. 0 or 1) into the fifth binary place.

(b) If the carry digit is 1, form the XS3 code of the units digit by adding 0011 to the least significant four digits of the sum of the two codes.

(c) If the carry digit is 0, proceed as in (b) but subtract 0011.

[2.21] The only choice arises for some codes where a 2-digit is needed. One solution is

| | | | |
|---|---|---|---|
| 0 | 0000 | 5 | 1011 |
| 1 | 0001 | 6 | 1100 |
| 2 | 0010 | 7 | 1101 |
| 3 | 0011 | 8 | 1110 |
| 4 | 0100 | 9 | 1111 |

[2.22] (a) Yes (b) No.

[2.23] (a)

| | | | |
|---|---|---|---|
| 0 | 0000 | 5 | 1110 |
| 1 | 1000 | 6 | 1010 |
| 2 | 1100 | 7 | 0010 |
| 3 | 0100 | 8 | 0011 |
| 4 | 0110 | 9 | 1011 |

(b) No.

(c) $H(N, N+1) = 1$; successive words differ from each other in only one binary position.

[2.24] A suitable algorithm for determining whether a code is uniquely decipherable is given by R. Ash, Information Theory, pp. 29-30.

Code A is not uniquely decipherable. An ambiguous sequence is 0001101011111000110.

Code B is uniquely decipherable.

[2.25] (a)

| | | |
|---|---|---|
| A | .5 | 1 |
| B | .2 | 01 |
| C | .1 | 0010 |
| D | .09 | 0011 |
| E | .06 | 0001 |
| F | .05 | 0000 |

Average length = 2.1 per group.

(b) 8, since $2^8 = 256 > 155 > 128 = 2^7$.

(c) Level 1 4 bits 14 characters (2 escape)

Level 2 8 bits 28 characters (4 escape)

Level 3 12 bits 60 characters (2 escape)

Level 4 16 bits remaining 53.

Average length $= 5\times.1\times4 + 9\times.02\times4 + 1\times.02\times8 + 20\times.005\times8$
$+ 7\times.003\times8 + 23\times.003\times12 + 37\times.0015\times12$
$+ 3\times.0015\times16 + 50\times.001\times16.$
$= 6.214.$

(d) Several solutions e.g.

Level 1 4 bits 5 characters (11 escape)

Level 2 8 bits 150 characters

Average length $= 5\times.1\times4 + .5\times8 = 6.$

## EXAMPLES 3

[3.1] 101 101 010 010 001 01, 100 101 010 001, 0.010 001 100 011 . . ., 0.000 101 100 010 . . .

[3.2] $(1531)_8$, $(216)_8$, $(0.6544)_8$, $(0.116)_8$,
$(359)_{16}$, $(8E)_{16}$, $(0.D64)_{16}$, $(0.27)_{16}$,
$(857)_{10}$, $(142)_{10}$, $(0.8369140625)_{10}$, $(0.15234375)_{10}$.

[3.3] 7.875, 1 101 111.000 111 000 110.

[3.4] For arithmetic on numbers of either sign, the only bit position which requires special consideration is the most significant one; for all the other positions the carry digit and the two digits of the numbers to be added have the same positive weight. In the most significant place the digits $(x_n, y_n)$ of the numbers have weight $-2^n$ while the carry digit

$c_n$ has weight $2^n$. Thus we have to consider the eight combinations below, of which the first two rows correspond to positive numbers ($x_n = y_n = 0$).

| $x_n$ | $y_n$ | $C_n$ | $S_n$ | |
|---|---|---|---|---|
| 0 | 0 | 0 | 0 | |
| 0 | 0 | 1 | 1 | (Overflow) |
| 0 | 1 | 0 | 1 | |
| 0 | 1 | 1 | 0 | |
| 1 | 0 | 0 | 1 | |
| 1 | 0 | 1 | 0 | |
| 1 | 1 | 0 | 0 | |
| 1 | 1 | 1 | 1 | |

The sum digits, $S_n$, are correct except when the register cannot hold the size of number resulting from the addition: such an overflow condition can be detected by the next carry digit, $c_{n+1}$, differing from the two sign digits, $x_n$, $y_n$.

[3.6] $$X = -2^{N-1}a_{N-1} + \sum_0^{N-2} 2^i a_i \quad \text{(Twos complement)}$$

$$X = -(2^{N-1}-1)a_{N-1} + \sum_0^{N-2} 2^i a_i \quad \text{(Ones complement)}$$

$$X = (-1)^{a_{N-1}} \cdot \sum_0^{N-2} 2^i a_i$$

[3.7]

| Twos comp. | Ones comp. | Sign and mod. |
|---|---|---|
| 101101(OV) | 101101(OV) | 001101(OV) |
| 001000(OV) | 001001(OV) | 101000 |
| 101100 | 101101 | 101100(OV) |
| 111111 | 111110 | 100001 |

[3.8] In the fixed point binary twos complement representation, all 48 bits could be 0 or 1, giving $2^{48}$ different combinations, each with a different integer value. In the floating binary representation the most significant bit is normalized to be 1; hence only $2^{47}$ different numbers are represented.

[3.9] The algorithm to calculate $g(x)$ using numbers in the floating binary representation could be such that there are (a) no, (b) one or (c) many representable $x$ for which $g(x)$ is computed as zero because of limitations on the accuracy of the representation and accumulation of errors in the calculation.

[3.10] $(153.451)_8$, $(179.529)_{10}$.

[3.12] (a) $0000$, $1\bar{1}0\bar{1}$, $\bar{1}0\bar{1}\bar{1}$

(b) Change 1s to $\bar{1}$s and vice versa.

(c) Floating 'ternary' would need four places for the exponent since $10^{40} \sim 3^k$ where $k = 40/\log_{10} 3 \sim 85 < 3^5$; 12 decimal places $12/\log 3 \sim 25$ ternary places i.e. a total of 29 places.

[3.13] A, the two forms of B ($B_1$ and $B_2$) and the results ($C_1$, $C_2$) need to be expressed in floating hexadecimal and their relation explained. $A = 16^3 \times (\cdot 2D2000)$, $B_1 = 16^\circ \times (\cdot DB3FFF)$ i.e. $B_1$ is nearly $16^\circ \times (\cdot DB4000)$.

$A + B_1 = 16^3 \times (\cdot 2D2DB3) = C_1 = 722.856201$.
$C_2 = 722.856445 = 16^3 \times (\cdot 2D2DB4)$.

One explanation is that $B_1$ is rounded down and $B_2$ rounded up; in the addition the numbers are brought to the same characteristic, $16^3$, and truncated i.e. not rounded. In the printing, $C_1$ and $C_2$ are rounded to 6D.

[3.14] $X = -2^{N-1}a_N + \sum_1^{N-1} 2^{i-1}a_i$; $-2^{N-1} \leq X \leq 2^{N-1} - 1$; $-167$.

[3.15] $16^{21} \times 37/128$; 54 800000; 55 520000.

[3.16] 1111 1110 1100 0000; 1010 0000 0000 0000.

$$\text{Value} = 2^{\left(\sum_{i=2}^{8} x_i 2^{8-i} - 64\right)} \left(-x_1 + \sum_9^{32} x_i 2^{-i+8}\right); \text{ about 8D.}$$

320.75 ≡ 0100 1001 1010 0000 0110 0000

-320.75 ≡ 1100 1001 0101 1111 1010 0000

[3.17] (a) $(1 - 2^{-24})16^{63} \geq |x| \geq 16^{-65}$ and $x = 0$ (b) $15.16^7$.

[3.18] $-275$, $-274$, $-237$; $1\bar{1}\bar{1}$, $\bar{1}1110$, $\bar{1}11110\bar{1}\bar{1}0\bar{1}$, $1\bar{1}1\bar{1}$; by the sign of the most significant digit; 167.

## EXAMPLES 4

[4.1] (a) $M(i, j) = BA + (n-1)i + j - 2.$

(b) $M(a, b, c, d)=BA+(r+q+1)\{(p-4)[(n+1)(a+1)+b]+(c-5)\}+d+q.$

[4.2] (a) $M(i, j) = BA + m(j-1) + i - 1.$

(b) $M(i, j, k) = BA + (m+2)[(n+1)(k+p) + j] + i + 1.$

(c) $M(i_1, i_2, i_3, \ldots i_n) = BA + ((\ldots((i_n-l_n)d_{n-1}+(i_{n-1}-l_{n-1}))$

$d_{n-2} + \ldots)d_2 + (i_2 - l_2))d_1 + i_1 - l_1$ where $d_i=u_i-l_i+1$ is the thickness of the ith dimension.

(d) $M(i, j) = BA + 3m(j-1) + 3(i-1).$

[4.3] $AT(1) = BA - 1$

$AT(i) = BA + 2i - 3$ for $i > 1.$

[4.4] Let the address of $A(1, 1) = BA$

$AT(1) = BA - 1$

$AT(2) = BA - 1 + (k+1)$

$AT(3) = BA - 1 + (k+1) + (k+2)$

$\vdots$

$AT(k) = BA - 1 + (k+1) + (k+2) + \ldots + (2k-1)$

$AT(k+1) = BA - 1 + (k+1) + (k+2) + \ldots + (2k-1)+2k=BA-1+\frac{k(3k+1)}{2}$

$AT(k+2) = BA - 1 + \frac{k(3k+1)}{2} + 2k$

$\vdots$

$AT(k+i) = BA - 1 + \frac{k(3k+1)}{2} + (i-1)2k$

$\vdots$

$AT(n-k+1) = BA - 1 + \frac{k(3k+1)}{2} + 2k(n-2k)$

$AT(n-k+2) = BA - 1 + \frac{k(3k+1)}{2} + (n-2k)2k+2k-1$

$\vdots$

$AT(n) = BA - 1 + \frac{k(3k+1)}{2} + (n-2k)2k+(2k-1)+ \ldots +(k+1)$

$= BA - 1 + 2nk - k^2 - k.$

[4.5] The contents of the access tables can be considered as pointers.

[4.6] We need the FORTRAN linear array

AVEC DIMENSION (N*(N+1)/2)

and A(I, J) → AVEC(I*(I-1)/2+J)

[4.7] Mapping function $M(i, j) = A_0 + n(i-1) - i(i-1)/2 + j - 1$.
Access table : store x as one-dimensional array $[1 : n\times(n+1)/2]$ and access $x[i, j]$ in location $AT[i] + j$ where

$$AT[1] = A_0 - 1, \quad AT[i] = (n-i/2) \times (i-1) - 1.$$

[4.9] Number of non-zero elements in $A + B = x + y$ - number of times non-zero elements in A and B occur in the same position.

For a given non-zero element in A, the prob. that that element in B is non-zero is $y/n^2$.

Hence, expected number of non-zero elements in $A + B$ is

$$x + y - xy/n^2$$

(ignoring the small probability that $a_{ij} + b_{ij} = 0$ if $a_{ij} \neq 0$).

If $A \times B = C$, $C_{ij} = \sum_{k=1}^{n} a_{ik}b_{kj}$.

The probability that a given term in this sum is non-zero is $xy/n^4$ and, ignoring exact cancellation as above, the probability that a given term $c_{ij} \neq 0$ is $xy/n^3$.

Hence the expected number of non-zero elements in $A \times B$ is $xy/n$.

[4.11] (a) Storage by rows.

Mapping function : $M(i, j) = BA + (N + 1)i + j$

Access Table : INTEGER ARRAY T(0::N);

$$T(i) = BA + (N + 1)i$$

(b) Storage by columns.

Mapping function : $M(i, j) = BA + (N + 1)j + i$

Access Table : INTEGER ARRAY T(0::N);

$$T(j) = BA + (N + 1)j$$

Access via $T(j) + i$.

Upper Hessenburg matrix.

Mapping Function : $M(i, j) = BA + (n+1)(i-1) - i(i - 1)/2 + j - 1$

Access Tables :

Store in a one dimension array X(1::(N*N+3*N-2)DIV 2);

Use access table : INTEGER ARRAY AT (1::N);

AT(i) = BA-1 + (i-1)(N+1-i/2)

General term found by accessing AT(i) + j.

[4.12] Possible algorithm to return the value of the element in the $r^{th}$ row and $s^{th}$ column.

```
x:= 0;
for I:= ROW(r) until (if r=n then m else Row (r+1)-1) do
if s= COLUMN(I) then begin X:= VALUE(I);
                           goto EXIT
                     end;
```

EXIT : Value required in X.

Insertion of a new element in $i^{th}$ row and $j^{th}$ column can be done by putting the whole of $i^{th}$ row at the end of the matrix with the value of ROW(i) changed to this new address. This implies moving the values already in the $i^{th}$ row to the end. The expected number of non-zero elements in $A^3$ is $p^3/n^2$.

[4.13] In row pointer/column index format

IQ : 1 4 6 8 9

IQN : 1 2 4 1 3 1 3 2

Pivot order : Either by column perm (2413) and row perm (4123).

Or by elements (4, 2), (1, 4), (2, 1), (3, 3).

Fill-in is 0, multiplication count is 3.

For natural order the fill-in is 6.

Matrix (ii) is structurally singular and has a transversal of length 5.

Column perm (1257436) puts non-zeros on the diagonal of (i).

## EXAMPLES 5

[5.1] (a) 13 stacks or queues initially with 4 items in each.

(b) The pack in Pontoon is like a queue, the cards being put back onto the bottom although not necessarily in the order they came off the top. The pack is not normally shuffled.

(c) Solitaire patience is more difficult as we require access to the middle of a list. Thus four sequentially stored arrays with room for 13 cards would seem appropriate.

[5.2] (a) A queue; it is almost impossible in most cafeteria to 'jump' the queue.

(b) Those waiting at the bus stop usually form a queue (except on the Continent) but those alighting from the bus do not do so in the order in which they entered it. The number of persons on the bus is limited; a one-dimensional array could be used.

[5.3] (a) Stack, Queue, Deque. (b) Stack, Deque. (c, d, e) Deque. (f) None.

[5.4] (a, b) Scroll, output restricted deque. (c) Scroll. (d) Output restricted deque. (e, f) None.

[5.6] (a) Five stacks with 14 locations and two with 15.

(b) First stack 23 last stack 22 other five stacks 11 locations each.

[5.7] A good algorithm will use a loop which links the head of the list as it currently stands to the previous head of list and prepares to iterate.

[5.8] The algorithm should take care to set the rear pointer when loading into an empty deque.

[5.9] Note the special cases of an empty circular list and one with just one node.

[5.10] Altogether this question requires twelve algorithms and to illustrate the type of algorithms required those for a linked queue (fig. 5.11) are given below.

(i) Let the lists to be joined have pointers (F1, R1) (F2, R2) and put the result in the list with pointers (F, R):

```
        if  F1 = Λ  then begin  F:= F2;  R:= R2;
                                goto  FINISH
                     end;
        F:= F1;  P:= F1;
LOOP: if  LINK(P) ≠ Λ  then begin  P:= LINK(P);
                                   goto  LOOP
                              end
                     else  LINK(P):= F2;
        if  R2 ≠ Λ  then  R:= R2  else  R:= R1;
FINISH:
```

(ii) Let original list have pointers (F, R) and suppose we wish to split the list just before node X into two lists with pointers (F1, R1) and (F2, R2):

```
        if  F = X  then begin  F1:= Λ;  R1:= Λ;
                               F2:= F;  R2:= R;  goto FINISH
                     end;
              F1:= F;
LOOP: if  LINK (F) ≠ X  then begin  F:= LINK (F);  goto  LOOP
                             end;
        F2:= X;  R2:= R;
        R1:= F;
```

This algorithm assumes there is a node X in the list; we could modify the algorithm to test if LINK (F) = Λ in which case the first list (F1, R1) is the whole list and the other list is null.

(iii) Deleting the list with pointers (F, R) and returning its nodes to the general pool of storage.

```
LOOP: if  F ≠ Λ  then begin
                               STACK (F, S);
                               F:= LINK (F);  goto  LOOP
                     end;
```

[5.11] The answer should be symmetric with respect to left and right links, to the algorithm given in the main text for inserting on the left.

[5.13] (i) 1 3 5 4 2 Let S represent a stacking operation and U an unstacking operation. Then the sequence for this permutation is S U S S U S S U U U.

(ii) 5 1 3 2 4 not obtainable on a stack or a deque.

(iii) 3 2 1 5 4 is obtainable on a stack by the sequence S S S U U U S S U U.

(iv) 2 3 5 1 4 not obtainable on a stack. Obtainable on a deque by sequence S S U S U S S S U U* U where U* is taking it off the bottom of the deque.

[5.20] There are two basic rules in stack manipulation:

Rule 1: if $a < b$ and $p_a < p_b$ then $p_a$ must be taken off the stack before $p_b$ is put on.

Rule 2: if $a < b$ and $p_a > p_b$ then $p_a$ must remain on the stack until after $p_b$ is put on.

Now if $i < j$ and $p_i < p_j$ then by Rule 1 $p_i$ is off the stack before $p_j$ is put on.

Also $k < i$ and $p_k > p_i$ then by Rule 2 we leave $p_i$ on the stack until after $p_k$ is put on.

These two statements mean that $p_k$ is removed from the stack before $p_j$ is put on so it is impossible for $p_k$ to come before $p_j$ in the output stream.

Example with $n = 5$

| 5 | 4 | 1 | 2 | 3 |
|---|---|---|---|---|
| $p_1$ | $p_2$ | $p_3$ | $p_4$ | $p_5$ |

if $k = 2$ $i = 4$ $j = 5$ then $k < i < j$

and $p_i < p_j < p_k$

$2 < 3 < 4$

```
[5.21] record NODE (integer DEGREE; real COEFF;
                                  reference (NODE) LINK);
reference (NODE) procedure POLYADD (reference (NODE) value P1, P2);
                 if P1 = NULL then P2 else if P2 = NULL then P1
                 else if DEGREE(P1) < DEGREE(P2) then
                   NODE(DEGREE(P1), COEFF(P1), POLYADD(LINK(P1), P2))
                 else if DEGREE(P1) > DEGREE(P2) then
                   NODE(DEGREE(P2), COEFF(P2), POLYADD(P1,LINK(P2)))
                 else begin real X; X:= COEFF(P1)+COEFF(P2);
                              if X≠0 then NODE(DEGREE(P1), X,
                        POLYADD(LINK(P1), LINK(P2)) else
                           POLYADD(LINK(P1), LINK(P2))
                    end;
```

This procedure creates a new list; the problem can be done iteratively as well as recursively.

## EXAMPLES 6

[6.1] One proof is by induction. When we consider a tree with n nodes (having assumed it true for a tree of n - 1 nodes) the crucial point is to show there is at least one node with only one line attached to it.

[6.2] (a) 9. (b) nephew-uncle; great uncle. (c) A degree 4, level 1; N 2, 4; R 0, 5:

[6.3]

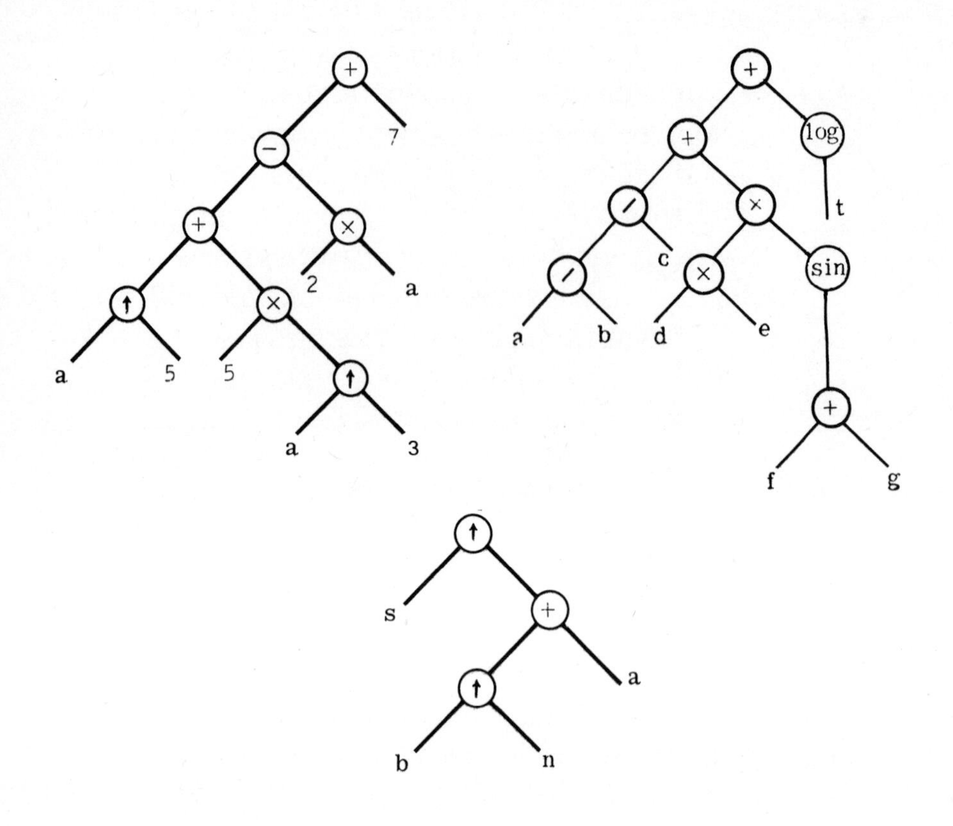


[6.4] The root of the tree is the first weighing of three coins against three coins. There are three branches which depend on whether the first three coins are greater, equal to or less than the other three.

[6.5] (a) R S T U V W J K L M N P F G H I B C D E A

(b) J R S T U K F G B C L M V W N H D P I E A

(c) A B C D E I P H L M N V W F G J K R S T U

Reverse endorder has only been defined for binary trees.

[6.6]

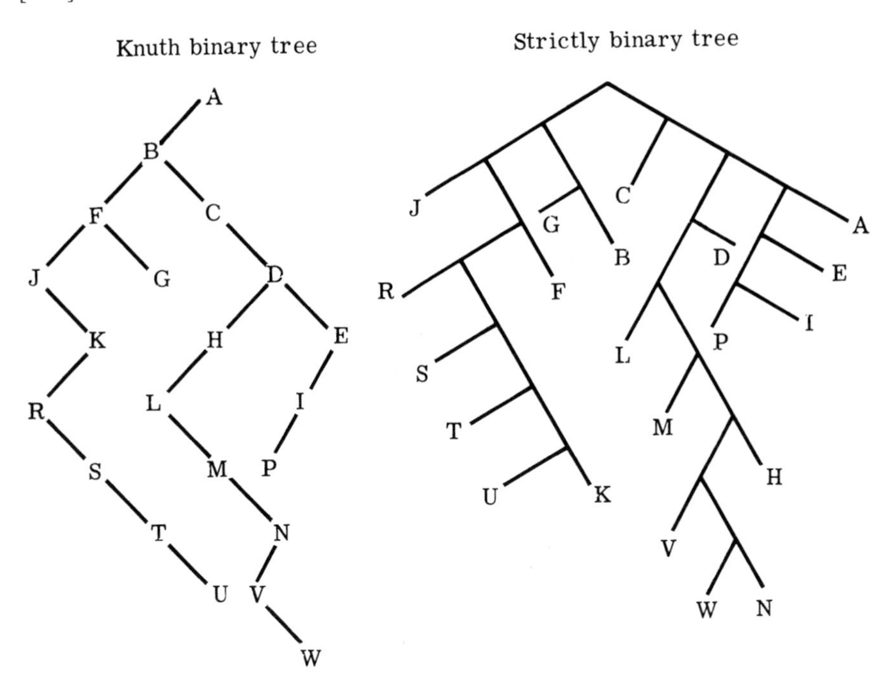


[6.7] A B F J K R S T U G C D H L M N V W E I P

same in both cases.

[6.9] 1111010100101110110000001100110111100000

[6.10] Numbering A → W 1 → 21

| Node no. | 1 | 2 | 3 | 4 | 5 | 6 | 7 | 8 | 9 | 10 | 11 | 12 | 13 | 14 | 15 | 16 | 17 | 18 | 19 | 20 | 21 |
|---|---|---|---|---|---|---|---|---|---|---|---|---|---|---|---|---|---|---|---|---|---|
| Node | A | B | C | D | E | F | G | H | I | J | K | L | M | N | P | R | S | T | U | V | W |
| FSON | 2 | 6 | * | 8 | 9 | 10 | * | 12 | 15 | * | 16 | * | * | 20 | * | * | * | * | * | * | * |
| Tag | + | - | - | - | + | - | + | + | + | - | + | - | - | + | + | - | - | - | + | - | + |
| BORF | 1 | 3 | 4 | 5 | 1 | 7 | 2 | 4 | 5 | 11 | 6 | 13 | 14 | 8 | 9 | 17 | 18 | 19 | 11 | 21 | 14 |

[6.12] See section 7.2.2 on analysis of the binary search algorithm.

If $n = 2^{i+1} - 1$ Path length $= (n+1) \log(n+1) - 2n$

[6.14] (i) (ii) (p, (q, r), (s, (t, u)))

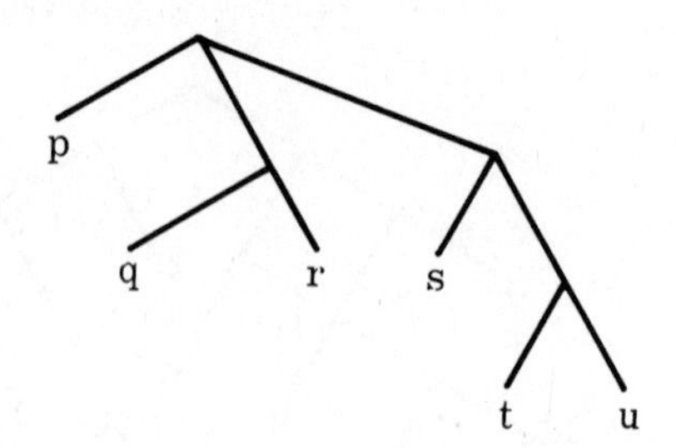


(a)

| | | | |
|---|---|---|---|
| 1 | 1, | 2, | 0 |
| 2 | 0, | p, | 3 |
| 3 | 1, | 4, | 6 |
| 4 | 0, | q, | 5 |
| 5 | 2, | r, | (11) |
| 6 | 3, | 7, | (12) |
| 7 | 0, | s, | 8 |
| 8 | 3, | 9, | (13) |
| 9 | 0, | t, | 10 |
| 10 | 2, | u, | (14) |
| 11 | 4, | 1, | 3 |
| 12 | 4, | 1, | 1 |
| 13 | 4, | 1, | 6 |
| 14 | 4, | 1, | 8 |

(b) Scan the entries from 1 to L. Whenever a type 2 or 3 entry is met say 2, m, n replace it by 2, m, s + 1 and add a new entry 4, 1, n at S + 1, increase S by 1. (S starts at L + 1 the first empty node.)

[6.16] (i) Polish expression - * a b / c d

A non-binary tree would be more convenient for (iv).

[6.21]

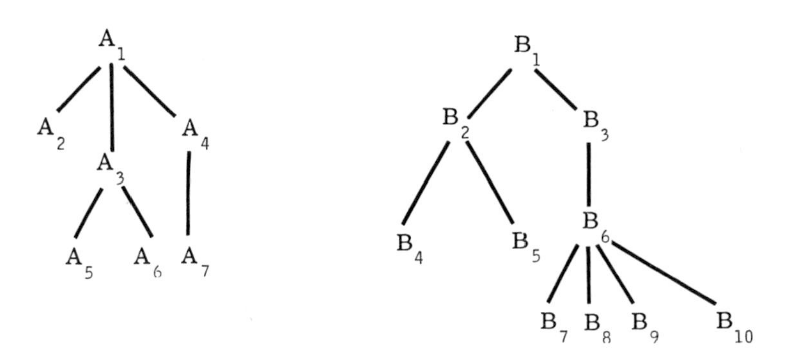


$(A_1(A_2, A_3(A_5, A_6), A_4(A_7))$, $(B_1(B_2(B_4, B_5), B_3(B_6(B_7, B_8, B_9, B_{10}))))$

Knuth binary tree for this forest is

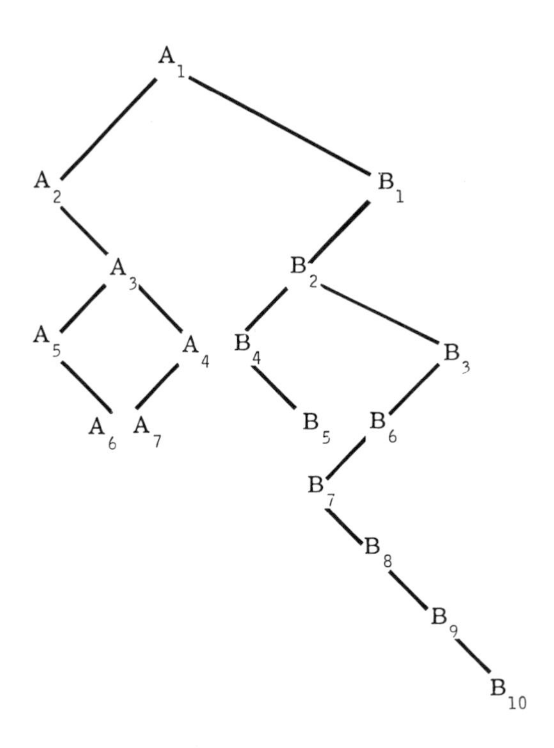

Preorder $A_1\ A_2\ A_3\ A_5\ A_6\ A_4\ A_7\ B_1\ B_2\ B_4\ B_5\ B_3\ B_6\ B_7\ B_8\ B_9\ B_{10}$

Endorder is defined as (1) Do nothing if the tree is empty
(2) Traverse the left subtree in endorder
(3) Traverse the right subtree in endorder
(4) Visit the root.

$A_6\ A_5\ A_7\ A_4\ A_3\ A_2\ B_5\ B_4\ B_{10}\ B_9\ B_8\ B_7\ B_6\ B_3\ B_2\ B_1\ A_1$

[6.22] The terminating binary sequence is

1110101111000110000011000100

## EXAMPLES 7

[7.2]

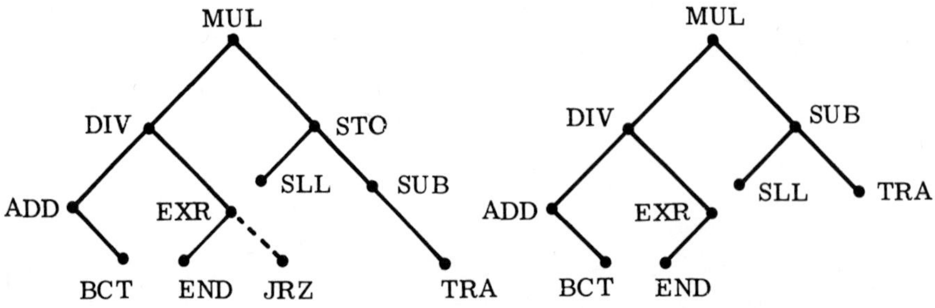

MUL
DIV
STO
ADD
EXR
SLL
SUB
BCT
END
JRZ
TRA

MUL
DIV
SUB
ADD
EXR
SLL
TRA
BCT
END

[7.3] 14.

Each binary sequence search tree of items is composed of a subtree including the root of the tree of r items which correspond to the r items before the item with key n and a subtree of the remaining items, rooted on n, which is attached to the rightmost element of the first subtree. Different search trees are obtained for each different subtree and position of the item n. Hence, if the number is $t_n$ we must have

$$t_n = t_0 t_{n-1} + t_1 t_{n-2} + \ldots + t_r t_{n-r-1} + \ldots + t_{n-1} t_0$$

where $t_0 = 1$. Recurrence relations of this form can be solved by using a generating function: see e.g. Riordan, Combinatorial Analysis.

[7.4] The offsets $r_i$ are 6, 7, 4, 5, 2, 3, 1.

[7.5] A deletion algorithm for the first method given in section 7.2.3 could be as follows.

```
begin integer  i, j, k;
      comment AR is the address of the root and we assume
      the node  x  being searched for is in the tree;
      i := j := AR;
      comment test to see if root is node to be deleted;
      if  x = key [AR] then begin  k := rlink [AR];
                                   if  k = 0  then  AR := llink [AR]
                                   else begin AR := k;
                                     while llink [k] ≠ 0 do k := llink [k];
                                     llink [k] := llink [j];
                                    end
                              end
else begin while x ≠ key [j] do
         begin i := j;  j := if  x < key [j]  then llink [j]
                                              else rlink [j]
         end;
         comment now we delete the node  j  whose father is node  i;
         k := rlink [j];
         if  k ≠ 0  then
              begin if rlink [i] = j then rlink [i] := k
                                     else llink [i] := k;
                  while llink [k] ≠ 0 do k := llink [k];
                  llink [k] := llink [j];
              end
         else if rlink [i] = j then rlink [i] := llink [j]
                               else llink [i] := llink [j];
      end
end Deletion Algorithm;
```

[7.6] Such a binary search tree divides into

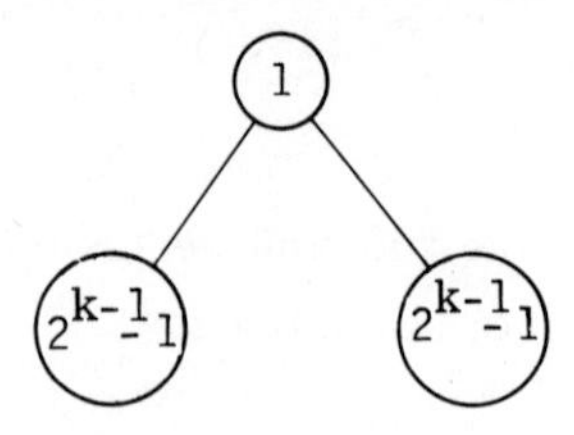


$$u_k = 1 + 2(u_{k-1} + 2^{k-1} - 1)$$

The first term is for the root and the second term for the two subtrees, each node of which is one level lower than it would be in $u_{k-1}$.

The solution with $u_1 = 1$ is $u_k = (k - 1)2^k + 1$, or since $n = 2^k - 1$, $k = \log_2(n + 1)$

$$u_k = (n + 1)\log_2(n + 1) - n.$$

[7.7] The algorithms for insertion and deletion are simplified by using two counters, one pointing to the end of the list with hash bit 0 and the other to the end of the list with hash bit 1. The two lists grow towards each other.

Deletion is most efficiently done by putting the last item in the appropriate list on top of the item to be deleted.

Total number of probes needed to create the table with $r$ hash bits of 0 and $(n - r)$ of 1 is $T = (1+2+\ldots+r) + (1+2+\ldots+\overline{n-r})$

$$= [r(r+1) + (n-r)(n-r+1)]/2.$$

Since $r$ is a binomial variable, $E\{r\} = np$, $E\{r^2\} = npq + n^2p^2$. The average number of probes for one item is $E\{T\}/n$ and substitution yields the required result on neglecting all terms of lower order than $n$.

[7.8] Let the average number of tests for $n$ items be $u_n$. Then $u_{2n+1} = \frac{1}{2n+1} \cdot 1 + \frac{2n}{2n+1}(1 + u_n)$. Since the probability is $\frac{1}{2n+1}$ that the search ends at the first test, and otherwise proceeds on a list of $n$ items.

Similarly $U_{2n} = \frac{1}{2n} \cdot 1 + \frac{(n-1)}{2n}(1 + u_{n-1}) + \frac{n}{2n}(1 + u_n)$.

Consider $u_{2^m-1}$ i.e. $u_n$ for $n = 2^m - 1$ and write

$t_m = (2^m - 1)u_{2^m-1}$ then $t_m = 2t_{m-1} + 2^m - 1$ which has the solution
$t_m = 2^m(m - 1 + \frac{1}{2^m})$.

Hence $u_{2^m-1} = m - 1 + \frac{1}{2^m} + \frac{1}{2^m-1}(m - 1 + \frac{1}{2^m})$.

Therefore $u_{2^m} \geq u_{2^m-1} \geq m - 1$ and since $u_n$ is increasing with $n$, the result follows.

[7.9]

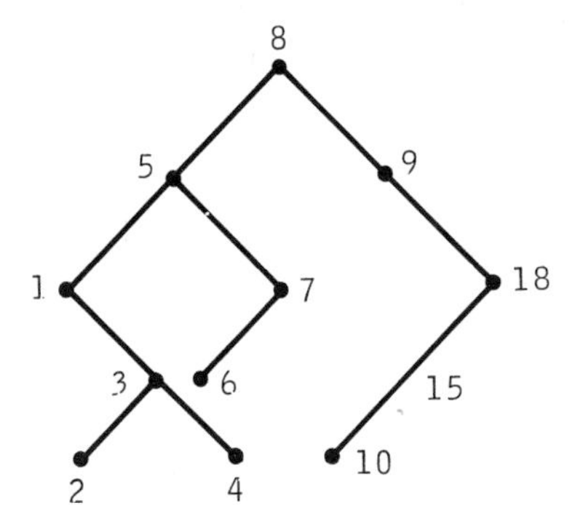


The Hibbard-Knuth deletion algorithm gives the resulting tree

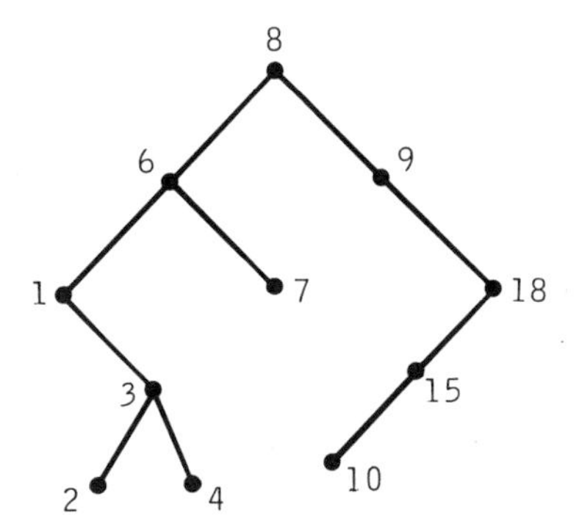

Since, out of (n + 1) items, an item is in the tree of n items with probability $\frac{n}{n+1}$ and not in this tree with probability $\frac{1}{n+1}$, we have

$$M^*(n+1) = \frac{n}{n+1} M^*(n) + \frac{1}{n+1} \overline{M}(n) .$$

By subtraction of the equations for $M^*(n+1)$ and $M^*(n)$

$$(n+1)M^*(n+1) - 2nM^*(n) + (n-1)M^*(n-1) = \overline{M}(n) - \overline{M}(n-1) = \frac{2}{n+1} .$$

By successive addition $(n+1)M^*(n+1) - nM^*(n) = 2H_{n+1} - 1$ and hence, similarly $nM^*(n) = 2\sum_1^n H_r - n$.

$$\text{Now } \sum_1^{n-1} H_r = \frac{1}{1} + \left(\frac{1}{1} + \frac{1}{2}\right) + \left(\frac{1}{1} + \frac{1}{2} + \frac{1}{3}\right) + \ldots + \left(\frac{1}{1} + \frac{1}{2} + \ldots + \frac{1}{n-1}\right)$$

$$= \sum_{r=1}^{n-1} \frac{n-r}{r} = \sum_{r=1}^{n} \frac{n-r}{r}$$

$$= n(H_n - 1) \text{ which gives the required result.}$$

For 10 keys, best possible sequences lead to trees of least longest path, corresponding to nearly binary divisions e. g. 5 3 1 2 4 7 6 9 8 10 with average length 2. 9.

The worst sequence gives the longest path e. g. 1 2 3 4 5 6 7 8 9 10 with average 5. 5.

[7. 12] (a) $(n + 1)/2$ successful, n unsuccessful

(b) $(n + 1)/2$ successful and unsuccessful

(c) $\sum_{i=1}^{n} ip_i$ successful, n unsuccessful.

Average number of comparisons for table $(\overline{C})$

$$= \frac{1}{M} \{\text{Comparisons for those in the table} + N \times \text{rest}\}$$

$$\overline{C} = \frac{1}{M} \{1 + 2 + \ldots + N + N(M - N)\} = N(2M - N + 1)/2M;$$

if $M = 60$, $N = 40$ then $\overline{C} = 27$.

[7. 13] Binary search tree depends on the algorithm. The algorithm 'binary search' in section 7. 2. 2 would give the tree

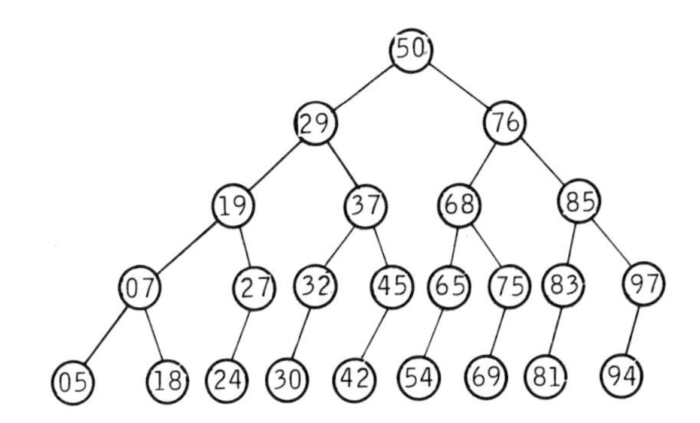


[7.14]

```
record NODE (reference (NODE) LLINK, RLINK; integer DATA, COUNT);
reference (NODE) Tree;
procedure SEARCH_INSERT (reference (NODE) value T; integer value X);
if T ≠ NULL then
    begin if DATA(T) = X then COUNT(T) := COUNT(T) + 1
        else if DATA(T) > X then
            begin if LLINK(T) ≠ NULL
                then SEARCH_INSERT(LLINK(T), X)
                else LLINK(T) := NODE(NULL, NULL, X, 1)
            end
        else if RLINK(T) ≠ NULL
            then SEARCH_INSERT(RLINK(T), X)
            else RLINK(T) := NODE(NULL, NULL, X, 1)
    end
else Tree := NODE(NULL, NULL, X, 1);
procedure MAXCOUNT (reference (NODE) value T; reference
                (NODE) result R; integer value result MAX);
comment MAX starts the procedure 0;
if T ≠ NULL then
begin if COUNT(T) > MAX then begin R := T; MAX := COUNT(T);
                                   end;
        MAXCOUNT (LLINK(T), R, MAX);
        MAXCOUNT (RLINK(T), R, MAX);
end;
```

The algorithm does a tree traversal and the largest value of COUNT is in the node referenced by R.

## EXAMPLES 8

[8.1] For $n$ random keys, the probability that the first selection pass requires no key interchange is the probability that the first key is the smallest i.e. $\frac{1}{n}$. Similarly, the second pass requires no interchange if what is now the second key is the smallest of the remaining $(n - 1)$ keys i.e. $1/(n - 1)$; and so on.

Hence the expected number of interchanges is

$$n - 1 - \sum_{i=2}^{n} \frac{1}{i} .$$

[8.3] Consider inserting the $(i+1)$st key into the list of $i$ sorted keys; there are $(i + 1)$ positions between the keys and at the ends where the new key can be inserted. Average number of comparisons to insert $(i+1)$st key

$$= [1 + 2 + 3 + \ldots + i - 1 + i + i]/(i + 1) .$$

Summing this over all the $(n - 1)$ passes gives

$$\frac{n^2+3n}{4} - \sum_{k=1}^{n} \frac{1}{k} .$$

The average number of interchanges is similarly analysed giving $(n^2 - n)/4$.

[8.4] First use the binary search algorithm of the type given in section 7.2.2 to find the correct place to insert the new key, then move keys down to make space to insert the new key.

[8.5] Bubblesort: 27 comparisons, 14 exchanges; Insertion: 19, 14; Insertion with binary search: 16, 14.

[8.6] The basic six steps for the natural two-way merge algorithm are given in section 8.3.2. We have to decide for step (5) what to do when either input list (A or B) is exhausted and there is still keys on the other. Some authors suggest copying the remainder of the unexhausted list onto one of the output lists but this can be slow and lead to difficulties. It is therefore suggested that the unexhausted list is merged as ordered strings onto the output list C and D. A useful subprocedure is one which inserts the current key z onto the appropriate output list, records this last value for comparison purposes later and increments the list which supplied z. It can also check for an exhausted list.

[8.7] Of the $\binom{l+m}{l}$ possible divisions (all assumed equally likely) into the $l$-string and m-string the smallest item will be in the $l$-string in $\binom{l+m-1}{l-1}$ cases i.e. a fraction $l/(l+m)$. If the smallest item is in the $l$-string, the first comparison leaves an $(l-1)$-string and an m-string to be merged. Similarly for the m-string containing the smallest item.

Hence $A(l, m) = \frac{l}{l+m}[1+A(l-1, m)] + \frac{m}{l+m}[1+A(l, m-1)]$,

which has the given formula for $A(l, m)$ as solution.

The number of cases where the r largest items belong to the $l$-string and the next largest to the m-string is $\binom{l+m-r-1}{l-r}$, and the probability is thus $\binom{l+m-r-1}{l-r}/\binom{l+m}{l}$.

[8.8]

| | | | | |
|---|---|---|---|---|
| 0 | 0 | 0 | 0 | 1 |
| 1 | 1 | 1 | 1 | 1 |
| 1 | 2 | 2 | 2 | 2 |
| 2 | 3 | 4 | 4 | 4 |
| 4 | 6 | 7 | 8 | 8 |
| 8 | 12 | 14 | 15 | 16 |
| 16 | 24 | 28 | 30 | 31 |
| 31 | 47 | 55 | 59 | 61 |

[8.9] Consider A = 1, B = 2, ... Z = 26 and construct the integer, N, in the scale of 26 represented by the first 4 characters of the word.

Try depositing the item represented by key N at $15 + 170N/26^4$ to leave, say 15 locations at each end initially and see what clashes occur.

[8.13] Note that after the initial adjustment phase we have a sequence

```
Odd    Odd    Odd   .....  Odd
Odd    Even   Even  .....  Even
Even   Odd    Even  ...... Even
Even   Even   Odd   ...... Even
 :      :      :            :
Even   Even   Even  ...... Odd
Odd    Odd    ............ Odd
```

Thus in case (a) we have a sequence that put us back to 11 ... 1 so we merge onto the empty tape.

[8.14]
```
procedure OETSORT (integer array k; integer value n);
    comment this procedure sorts the n keys
    k(1) ... k(n).  Logical variable L is true for
    first half stage and false for second half stage.
    FLAG makes the decision to exit from the
    procedure if there is no key exchange during a
    pass (two half stages);
begin logical L, FLAG;
                FLAG := L := false;
    while L or FLAG do
    begin L := not L;  if L then FLAG := false;
          for i := (if L then 2 else 3) step 2 until n do
          if k(i - 1) > k(i) then
              begin Exchange (k(i - 1), k(i)); FLAG := true
              end;
    end
end OETSORT;
```

An alternative stopping condition is:- (B). Stop when a half-stage occurs without an exchange, but we must do a minimum of two half-stages.

For the 10 keys OETSORT requires 25 exchanges and 45 comparisons (41 comparisons if stopping condition (B) is used). Bubblesort requires 25 exchanges and 42 comparisons.

[8.15] Funnelsort 23 exchanges, 39 comparisons.
Bubblesort 23 exchanges, 45 comparisons.

[8.16] Let the keys start in array k and finish in array sort. Set array count to 1 initially

```
for i := 1 until N do
for j := i + 1 until N do
if k(i) < k(j) then count (j) := count(j) + 1
                else count(i) := count(i) + 1;
for i := 1 until N do
Sort (count(i)) := k(i);
```

Number of key comparisons = $N(N - 1)/2$
For the nine keys this gives 36 comparisons.
Bubblesort: 30 comparisons, 14 key exchanges.

[8.17] By standard Quicksort 27 comparisons
20 key movements (12 + 8 insertions of ranked key).

In proving the extra storage space is of the order $\log_2 n$ locations consider the worst case when the lists always divide into two equal halves.

# Index

The items marked with an * appear frequently throughout the book and only the more important references to them are recorded.


Access tables (see also Arrays) 56-61, 66-9, 243-5

Accessing* 51, 62-4, 68, 70, 98-9, 194

Address* (see Link)

Aho, A. V. 203-4, 208, 232

ALGOL* 20, 39, 50-1, 68, 77, 99, 102-4, 150, 152

ALGOL 68 51, 154

ALGOL W 51, 68, 72, 87, 99, 106, 190, 221-2, 249, 259

Anderson, Nils (ix)

APL 5, 183, 223

Arden, B. W. 183-4

Arithmetic expressions as trees 114, 119-20, 149, 152, 157-8, 250, 252

Arrays* (see also Access Tables and Mapping Functions) 50-73, 113-4, 243-5

- banded matrix 66, 73
- Hessenburg 69, 244
- jagged 59, 67
- sparse 61-73, 182-3, 244-5
- symmetric or skew-symmetric 56, 73
- triangular 56-9, 67
- tridiagonal 66
- two-dimensional (matrix)* 52-4, 74, 113-4, 182

Ash, R. 25, 239

Assemblers 104, 166, 182

B50000 75

Balanced ternary 45-6, 48-9

Base Address* 51, 54-6, 79

Bell, J. R. 183-4, 191

Berztiss, A. T. 65-6, 99-100, 147-8

Binary coded decimal 36

Binary element 8, 33

Binary fraction 35, 38

Binary integer 11, 34, 37, 43-8

Binary search 132, 163-9, 183-9, 204, 251, 254, 256, 258-61

Binary sequence search tree (BSST) 169-73, 183-90, 204, 206, 254-9

Binary trees* (see also Traversing trees and Tree representation) 112-3, 118-32, 143-7, 150-9

- binary search tree 167, 256
- complete binary tree 131, 145, 150, 167, 210
- Huffman coding tree 21-2
- Knuth binary tree 113, 122-4, 143, 251-3
- sorting using binary trees 204-11, 223, 233
- strictly binary tree 113, 124-7, 135-8, 143, 251

Bits* 18, 19, 32, 39, 51
Blank tape 10
Boothroyd, J. 224-5, 234
Braden, R. T. 147-8
Brent, R. P. 191
Brooks, F. P. 42
Bruno, J. 183-4, 191
Bytes 32-3, 36-46, 51

Characteristic (see exponent)
Characters* 4, 5, 9-12, 32
Cheatham, T. E. 147-8
Circular list (see Linear list)
Clustering 175-6
  primary 176
  secondary 177, 182
COBOL 50
Code words 12-22, 26-30
Codes*
  ASCII 36
  biquinary 23-4, 30
  block 16, 18, 26, 28, 31, 236-7
  compact 20, 28-9
  EBCDIC 36
  genetic 29
  Gray 30
  instantaneous 20, 28
  non-singular 20
  optimal 20, 26, 29
  uniquely decodable (deciperable) 20, 30-1, 239
  weighted 11, 22-4, 30
  XS3 23, 30, 239
Coding theory (information theory) 16-27, 163, 195-6
Coffman, E. G. 173, 183-4, 191
Collision of keys (see Key transformation)
Combining lists 52, 94, 99, 102, 105-6, 246-9
Compilers 46, 56, 114, 173, 182, 203, 221
COMIT 99
Conversion between scales 33-6, 43-5
Copying lists 52
Core storage 61, 194, 196
Curtis, A. R. 73

de Buchet, J. 72
Decimal fractions 35
Decimal integers 34
  packed 36, 45
Decoding 14-17, 20, 236
Degree of a node (see Trees)
Deleting* 51, 63-5, 74-5, 79-81, 87-102, 162, 170-2, 175-9, 185-9, 255-7
Deques 74-5, 78, 80-1, 87-92, 100-2, 246-8
de Villiers, E. v. de S. 73
Dictionary 107, 132, 152
Dijkstra, E. W. 75, 78
Dodd, M. 191-2
Dope vector 53-4, 56
Doubly linked list (see Linear list)
Double length representation 42
Duff, I. S. 72

Elements* (see Nodes)
Elson, M. 65-6, 99-100, 147-8
End-around carry 38
Entropy 19-21, 26-9, 238
Erase 10, 12, 25, 235

Error correction 12-16, 26-30, 236-7

Error detection 10, 12-16, 25-30, 236-7

Escape characters 31, 240

Exponent 39-42, 242

Extension of message sources 20, 238

Feurzeig, W. 224-5

Fields* 50

FIFO (see Queue)

File 188, 193-4

Fixed point 38, 44, 47, 241

Floating point 39-48, 241-2

Flores, I. 199, 223-4, 232-4

Floyd, R. W. 208, 223-4

Forests (see also Trees) 112, 122-4, 135, 154-5, 253

Forsythe, A. I. 147-8

FORTRAN 46, 50, 53, 56, 67, 99, 244

Frazer, W. D. 223, 225, 233

Friend, E. H. 223-4, 232

Galler, B. 183-4

Garwick, J. V. 83-4, 107

Gilstad, R. L. 224-5

Graph 109, 112, 152

Gustavson, F. G. 72-3

Hamming distance 13-16, 25-27, 30, 235-6

Hash addressing or Hash coding (see Key transformation)

Hexadecimal 4, 33, 35, 39-47, 242

Hibbard, T. N. 172, 183-4, 223, 225, 257

Hildebrandt, P. 223, 225, 233

Hoare, C. A. R. 99-100, 223-4

Hopcroft, J. E. (See Aho)

Huffman, D. 20, 236

Huffman algorithm 20-2, 26, 28, 31, 115, 135, 146-7, 236-40

IBM machines 24, 32, 39-42, 46-8, 221

D'Imperio, M. E. 99, 147-8

Information, measure of 18, 163, 195

Information theory (see Coding theory)

Inserting* 51, 63-5, 68-70, 79-98, 101-2, 162, 169-70, 185-90

Integers, binary representation 37-8

IPL 99

Isbitz, H. 223, 225, 233

Issac, E. J. 224-5, 234

Items* (see Nodes)

Iverson, K. E. 5, 42, 147-8, 183-4, 223-4, 233

Jennings, A. 73

Johnson, L. R. 147-8, 183-4, 191

KDF9 12, 74

Key* 51, 57, 162, 193

thick or thin 193

Key comparisons

in searching 163-173

in sorting 195-204, 208-11, 214, 217, 222, 226, 229-32, 260-3

Key interchanges (or transfers or exchanges) 195, 197-217, 222, 226, 229-32, 260-3

Key transformation 162, 173-83

collision methods 175-82, 192

chaining 178-9, 181-3, 192
linear 176-7, 181-3, 192
modified linear 192
quadratic 177, 181-3, 192
quadratic quotient 177, 181-3, 192
random 178, 181-5, 192
hash addressing (hash coding) 61, 73, 174-92, 212
in address calculation sorting 211-4
overflow tables 176-9
scatter storage table (SST) 175-9, 182, 185-6, 191, 212
Keys, detached 60, 103, 162, 175, 193-4
Knuth, D. E. (vii), 42, 65-6, 73, 83-4, 89, 99-101, 107, 112-3, 116, 122, 147-8, 158, 172, 183, 185, 191, 223-4, 232-4
Korfhage, R. R. 147-8, 159-60
Kritzinger, P. S. 233

Letters* 3-5, 11, 32
Lewis, T. G. 184-5, 191
Lexicographical order 55-6, 66, 132
LIFO (see Stack)
Linear lists* 50, 74-108
circular (ring structures) 77, 81, 92-4, 102-3, 246
doubly linked 73, 94-7, 102
linked 64-5, 72, 74, 85-99, 101-6, 178, 192, 246
sequentially stored 78-84, 97-9
Link* 51
Linked list (see Linear list)
LISP 99
London, R. L. 208, 223-4
Lorin, H. 223-4, 232-4
Lum, V. Y. 191-2

McIlroy, M. D. 183-4, 191
McKellar, A. C. 223, 225
Magnetic discs or drums 61, 188, 194, 196
Magnetic tapes 61, 194, 196, 214, 218-21, 227-8
Mantissa 39-42, 44, 46
Mapping functions 54-6, 66-9, 243-4
Mark reader 29
Martin, W. A. 204, 223-4, 232-4
Matrix (see Arrays)
Maurer, W. D. 183-5, 191-2
Merging (see Sorting methods)
Message symbols* 14-20
Morris, R. 183-4, 191-2

Neivergelt, J. 184-5
Nested parenthesis 151, 155, 252-3
Nieder, A. (see Pooch)
Nodes* (see also Trees) 50
Normalisation 40-1

Octal 28, 33-4, 43-5
Ogbuobiri, E. C. 73
One's complement 30, 37, 43-8, 241
Open addressing 176
Operating system 103, 195, 221
Ordered data 195, 197, 199, 210, 216-7, 221-3, 232
Output restricted deque 101, 246
Overflow* 38, 76, 78-8, 82-5, 241

Packed decimal integers (see Decimal integers)
Pages and paging 33, 183, 195, 214
Paper tape codes* 9-12, 16, 22-8, 235
Parity 10-12, 23-6, 235
PASCAL 56
Path* 109, 132-5
Path length* 143-7, 168, 172, 187, 251, 258
  in a binary tree 143-7, 150
  minimum (shortest) 145, 152
  weighted (see also Huffman algorithm) 135, 145-7
Pegasus 11
Perlis, A. J. 65, 147-8
Peters, J. G. 233
Pierce, J. R. 25
PL/1 50-1, 53, 99, 154
Pointer* (see Link)
Polish notation (including Reverse Polish) 77, 120, 152, 157, 252
Pooch, U. W. 65-6, 72
Price, C. E. 120
Programming projects
  group 71-3, 191-2, 231-4
  individual 107-8, 156-61, 190-1
Propositional calculus (Boolean algebra) 158-60
Punched cards 9, 24-5

Queues (FIFO) 74-81, 87-104, 245-8

Ranking of keys 204-6, 208, 210
Record* 188, 193-4, 227
Recursion and recursive definitions 77, 110, 116-20, 130, 135, 204-6
Reference* (see Link)
Reid, J. K. 72-3
Ring structures (see Linear lists)
Riordan, J. 147-9, 254
Rose, D. R. 73
Row major order (see Lexicographical order)

Scale factors 38
Scales of notation 4, 33-6
Scanning (see also Searching) 162-73, 183
  controlled 163-73
  directed 163-4, 189, 196, 202-7, 252, 258
Scatter storage tables (see Key transformation)
Schay, C. 183-4
Scoins, H. I. 135, 147-8
Scowan, R. S. 224-5, 234
Scroll 101, 246
Searching* (see also Scanning, Key transformation and Traversing trees) 51, 132, 140, 162-92, 193
Separating lists 52, 99, 102, 247
Severance, D. G. 184-5
Shave, M. J. R. 65-6, 99-100, 147-8
Shell, D. A. 224-5, 234
Siewiorek, D. P. 65-6, 99-100, 147-8
Sign digit* 37, 39-40, 46, 141, 241
Sign and modulus (or sign and magnitude) 36, 39-48, 241-2
SIMULA 99
Singleton, R. C. 224-5, 234

SNOBOL 99

Sobel, S. 224-5

Sorting* 51, 60, 99, 103, 193-234, 260-5

Sorting methods (including merging methods)

- address calculation 211-4, 222-4, 226, 233-4, 261-2
- balanced two-way merge 214-8, 233
- bubblesort 199-202, 211, 221-2, 226, 229-31, 233, 261, 263
- counting 230-1, 263
- cubic selection 199, 232
- funnel 202, 229-30, 263
- heapsort 199, 208, 223-4, 232
- insertion 202-3, 221-2, 226, 233, 260
- merging 214-28, 233, 261-2
- monkey puzzle 206, 223
- multiway merge 218
- natural selection 233
- natural two-way merge 216-8, 221-2, 226, 233, 261
- odd-even transposition 202, 229, 262-3
- oscillating merge 221, 224
- pigeon-hole 195, 211, 228
- polyphase merge 218-21, 224-6, 228, 262
- quadratic selection 298-9, 222, 226, 232
- quicksort 204-11, 221-4, 231, 233, 263
- radix exchange 206-8, 223, 225, 233
- replacement selection 232-3
- sample sort 211, 224-5, 233
- selection 196-9, 201, 221-3, 226, 232-3, 260
- shellsort 224-5, 234
- treesort 199, 208, 223-4
- tournament 199, 223, 232

Sparse matrices (see Arrays)

Spruth, W. G. 183-4

Stacks (LIFO) 74-9, 82-8, 100-7, 127-30, 245-8

Stone, H. S. 65-6, 99-100, 147-8

Structures, Information or Data (see also Arrays, Linear lists and Trees) 50

Subroutines 76-7, 103

Subtrees* 110, 137-8, 142-3, 156

Symbols* 1-7, 9-31

Symbol tables 182

Table lookup 173, 184

Tag 130-1, 138-42, 150, 154-5, 251

Thread link 141-2, 151

Traversing trees 113-20

- binary tree traversal 118-20, 127-31, 150, 155
- endorder 155, 254
- family order 116-9, 150
- level by level (constant depth) 115, 119, 150
- preorder (Prefix walk) 116, 119, 140, 150, 154-5, 157, 254
- reverse endorder 135, 150, 250
- suffix walk 116, 119, 150
- symmetric order 120, 127-30, 150, 206

Trees* (see also Binary trees, Forest, Path) 50, 109-61

- branch node 112
- decision 114, 150, 250

degree of a node 112, 149, 249
equivalent 111
family 109, 112, 149, 249
free 112
labelled 132
length between nodes 112
level in a tree 112, 149, 249
ordered 111
rooted 109
similar 111
terminal node (leaf) 112
Tree representation 127-43
'above' representation 132-5, 146-7, 150
binary tree 127-32
terminating binary sequence (tbs) 135-8, 143, 150, 155-6, 251, 254
two links and a tag 138-41, 150, 251
Tree transformation 120-7
forests into Knuth binary trees 122-4, 150, 154-5, 251, 253
trees into strictly binary trees 124-7, 131, 135, 150, 156, 251
Two's complement 37-9, 43-8, 241-2

Ullman, J. D. (see Aho)
Underflow* 42, 76, 78, 83, 88, 90, 91, 93

Wegner, P. 99-100
Williams, J. W. J. 208, 223-4
Willoughby, R. A. 73
Wilson, L. B. 73
Windley, P. E. 223-4
Wong, C. K. 233
Words* 32-3, 37-9, 51

Yuen, P. S. T. 191-2

Zero memory source 19, 28